SpringerBriefs present concise summaries of cutting-edge research and practical applications across a wide spectrum of fields. Featuring compact volumes of 50 to 125 pages, the series covers a range of content from professional to academic.

Typical publications can be:

- A timely report of state-of-the art methods
- An introduction to or a manual for the application of mathematical or computer techniques
- A bridge between new research results, as published in journal articles
- A snapshot of a hot or emerging topic
- An in-depth case study
- A presentation of core concepts that students must understand in order to make independent contributions

SpringerBriefs are characterized by fast, global electronic dissemination, standard publishing contracts, standardized manuscript preparation and formatting guidelines, and expedited production schedules.

On the one hand, **SpringerBriefs in Applied Sciences and Technology** are devoted to the publication of fundamentals and applications within the different classical engineering disciplines as well as in interdisciplinary fields that recently emerged between these areas. On the other hand, as the boundary separating fundamental research and applied technology is more and more dissolving, this series is particularly open to trans-disciplinary topics between fundamental science and engineering.

Indexed by EI-Compendex, SCOPUS and Springerlink.

Rabiu Muazu Musa · Anwar P. P. Abdul Majeed ·
Aina Munirah Ab Rasid ·
Mohamad Razali Abdullah

Data Mining and Machine Learning in Sports

Success Metrics for Elite Goalkeepers in European Football Leagues

Springer

Rabiu Muazu Musa
Center for Fundamental and Continuing
Education
Universiti Malaysia Terengganu
Kuala Nerus, Terengganu, Malaysia

Anwar P. P. Abdul Majeed
School of Robotics, XJTLU Entrepreneur
College (Taicang)
Xi'an Jiaotong-Liverpool University
Suzhou, China

Aina Munirah Ab Rasid
Center for Fundamental and Continuing
Education
Universiti Malaysia Terengganu
Kuala Nerus, Terengganu, Malaysia

Mohamad Razali Abdullah
Faculty of Health Science
Universiti Sultan Zainal Abidin
Terengganu, Malaysia

ISSN 2191-530X   ISSN 2191-5318 (electronic)
SpringerBriefs in Applied Sciences and Technology
ISBN 978-981-99-7761-1   ISBN 978-981-99-7762-8 (eBook)
https://doi.org/10.1007/978-981-99-7762-8

This Springer imprint is published by the registered company Springer Nature Singapore Pte Ltd.
The registered company address is: 152 Beach Road, #21-01/04 Gateway East, Singapore 189721, Singapore

Paper in this product is recyclable.

Acknowledgement

We would like to express our gratitude to Dr. Zahari Taha for his guidance and valuable suggestions, which were instrumental in making this book a reality.

Rabiu Muazu Musa
Anwar P. P. Abdul Majeed
Aina Munirah Ab Rasid
Mohamad Razali Abdullah

Contents

Chapter 1
Recent Advancements in Data Mining and Machine Learning Applications in Evaluating Goalkeepers' Performances in Elite Football

Abstract In this chapter, we explored how evaluation methods have evolved to assess the performance of elite football goalkeepers and how data mining and machine learning have advanced the field of sports analytics. We also identified the key performance indicators that affect the success of goalkeepers in the European football championship. Furthermore, we described the data collection process, data sample and features, as well as various statistical analyses (both univariate and multivariate) that were used to accomplish the objectives of this study.

Keywords European football championship · Machine learning models · Clustering algorithms · Data mining · Performance evaluation · Goalkeepers' performance

1.1 An Overview of Current Analysis of Goalkeepers' Performance

The current sports analysis environment has experienced a substantial emphasis on the evaluation of goalkeepers' performance, with the implementation of data mining and machine learning techniques playing a critical role in shaping this trend [1]. This development is propelling a more advanced comprehension of goalkeepers' contributions to team achievement and is enabling data-oriented decision-making procedures in the realm of sports.

In recent times, the assessment of goalkeepers' performance has evolved beyond traditional metrics like saves and goals conceded. Instead, a more comprehensive approach that considers a diverse range of factors has been adopted. With the aid of advanced tracking technologies such as tracking systems and wearable devices, it is now possible to gather detailed data on various aspects of a goalkeeper's performance during matches [2, 3]. This includes information on their movements, positioning, distribution, and decision-making.

One key trend is the integration of spatial analysis, which involves the examination of goalkeeper positioning and coverage areas in relation to the goal area and the

© The Author(s), under exclusive license to Springer Nature Singapore Pte Ltd. 2024
R. M. Musa et al., *Data Mining and Machine Learning in Sports*,
SpringerBriefs in Applied Sciences and Technology,
https://doi.org/10.1007/978-981-99-7762-8_1

ball. This provides insights into the shot-stopping efficiency, anticipation skills, and overall tactical awareness of a goalkeeper [4–6]. Additionally, the use of heat maps and trajectory analyses aids in identifying patterns and vulnerabilities in goalkeepers' positioning and movement.

Data mining methodologies are currently being employed to extract valuable insights from extensive amounts of gathered data. To unveil the correlations between goalkeepers' actions and match outcomes, machine learning algorithms are utilized, facilitating informed decisions for coaches and analysts regarding training regimens, tactics, and game strategy [7–10]. Recent advancements in machine learning have resulted in the creation of predictive models that analyse the performance trends of goalkeepers and anticipate potential game-changing moments [11–13]. These models take into account various contextual factors, such as opponent characteristics, match conditions, and team dynamics, to offer real-time recommendations for optimizing goalkeeper decisions. In addition, data-driven benchmarking has become increasingly important in evaluating the performance of goalkeepers. Comparisons are made against both historical data and peers to identify areas of strength and areas that need improvement. This enables a more objective evaluation of the contributions of goalkeepers to the success of the team.

1.2   Recent Updates in Data Mining and Machine Learning Applications in Sports

The utilization of machine learning in sports has received considerable attention in recent times. This topic has been the subject of several recent review papers, indicating its increasing popularity [14–20]. In this section, we aim to provide the readers with an overview of how this technology is used in the field of sports for both the identification of essential variables relevant to specific sports as well as identification of talented athletes.

Tan et al. [21] conducted a study to examine the effectiveness of various machine learning models, including k-nearest neighbours (kNN), support vector machine (SVM), artificial neural networks (ANN), Naive Bayes (NB), and random forest (RF), in classifying different types of Sepak Takraw kicks. Data were collected from inertial measurement units (IMU) placed on the participants' shanks and extracted various statistical features from the IMU data. The data was then fed into the machine learning models for training and testing using a 70:30 holdout ratio. The study demonstrated that the ANN model achieved excellent classification performance, correctly identifying serve (a.k.a. "tekong"), feeder, and spike kicks without any misclassifications in the test dataset.

Thabtah et al. [22] investigated the effectiveness of machine learning model and its associated features in predicting the outcomes of National Basketball Association (NBA) games. They utilized NB, ANN, and decision tree (DT) models and obtained the NBA Finals Team Stats Dataset from 1980 through 2017. Three different feature selection techniques were employed: multiple regression, correlation feature set, and the RIPPER algorithm. The study indicated that defensive rebounds were a significant feature influencing the results of NBA games.

Echterhoff et al. [23] studied the classification of gait and jump in modern equestrian sports. They attached a smartphone to the horse saddle to capture accelerometry and gyroscopic data. A total of 268 features were extracted from the data, considering both the time and frequency domains. They investigated 22 variations of machine learning models using a ninefold cross-validation technique. The study demonstrated that the cubic-SVM achieved a classification accuracy of 95.4% in distinguishing the four evaluated classes: walk, trot, canter, and jump.

McGrath et al. [24] focused on the detection of cricket fast bowling using machine learning with IMU-acquired data. They evaluated different machine learning models, including linear SVM (LSVM), polynomial SVM, ANN, RF, and gradient boosting (XGB), using a reduced feature set of 223 features. They also investigated different sampling frequencies to assess the models' classification efficacy for bowling and non-bowling events. The study illustrated that the SVM models generally achieved slightly higher classification accuracy compared to other models, with all models surpassing 95% accuracy, indicating the significance of the extracted features.

Worsey et al. [25] explored the classification of boxing punches using machine learning and IMU data. The participants performed jab, cross, hook, and uppercut punches with both hands. Principal component analysis (PCA) was applied to reduce the dimensionality of the extracted features. Six machine learning models were evaluated: logistic regression (LR), LSVM, Gaussian SVM (GSVM), ANN, RF, and XGB. The models' hyperparameters were optimized using exhaustive grid search with fivefold cross-validation. The study demonstrated that the untuned GSVM and the ANN model performed well in classifying the punches.

Duki et al. [26] conducted a feature selection investigation in the classification of Taekwondo kicks. They extracted nine statistical features from IMU acceleration data, including minimum, maximum, mean, median, standard deviation, variance, skewness, kurtosis, and standard mean error. The significance of these features was assessed using ANOVA and chi-square tests. They fed the features into different machine learning models, namely SVM, RF, kNN, and NB, using a 60:40 holdout ratio. The study showed that the features identified through the ANOVA method achieved better classification accuracy, up to 86.7% using the RF model, compared to using all features or the features identified by other methods.

Abdullah et al. [27] employed six transfer learning models with an optimized SVM classifier to classify skateboarding tricks. They investigated two input image types extracted from an IMU fixed on the skateboard: stacked raw signals (RAW) and continuous wavelet transform (CWT). The SVM model's hyperparameters were optimized using grid search, and a 60:20:20 training, testing, and validation ratio was utilized. The study concluded that the CWT-MobileNet-Optimized SVM pipeline outperformed other combinations in terms of computational time and classification accuracy.

Musa et al. [28] examined the identification of high-performance volleyball players (HVP) and low-performance volleyball players (LVP) using machine learning based on anthropometric variables and psychological readiness indicators. They employed the Louvain clustering algorithm to distinguish player performance and utilized logistic regression for classification. Due to the skewed nature of the data, the authors employed the Synthetic Minority Oversampling TEchnique (SMOTE) to address overfitting concerns. The study demonstrated excellent identification of player classes based on the developed pipeline.

In summary, it is evident from the above literature that machine learning has gained significant traction in the sports domain and has shown promise in achieving accurate predictions in various cases, highlighting its invaluable contribution to sports in general.

1.3 Mann–Whitney U-Test

The Mann–Whitney U-test is a univariate mathematical technique predominantly utilized for comparing means. This test falls under the category of dependency test analysis, where the variables under consideration are categorized as independent or dependent [29]. The underlying rationale of the test revolves around the notion that variations in the average scores of the dependent variable(s) can largely be attributed to the direct influence(s) of the independent variable.

In contrast to the t-test and F-test, the Mann–Whitney U-test is categorized as a nonparametric analysis. This classification indicates that there is no pre-established assumption about the distribution means of the variables of interest in the population samples. Unlike the analogous one-way ANOVA, the nonparametric Mann–Whitney U-test does not make assumptions about the normal distribution of the underlying data [30].

The test has been effectively utilized across various sports and has demonstrated its efficacy in highlighting distinctions among two or more tiers of performance categories. For example, in a recent investigation, the Mann–Whitney U-test was employed to pinpoint technical and tactical performance indicators capable of distinguishing between successful and unsuccessful teams in elite beach soccer competitions [31]. In a preceding research endeavour, the Mann–Whitney U-test was also proficiently utilized to examine the likelihood of experiencing sports-related injuries among British athletes engaged in wheelchair racing [32].

1.4 Chi-Square Analysis

The primary purpose of a chi-square test is to facilitate the analysis of categorical data. Equally, the chi-square test operates on a dataset that has been categorized and counted. As a result, this test is not suitable for handling numeric data that is parametric or continuous. It is important to highlight that the dataset required for conducting the chi-square test should be presented in terms of frequencies, specifically count datasets, rather than using percentages, percentiles, or relative frequencies [33–36]. In the present study, the chi-square analysis is employed to investigate the relationship between variables that are categorical or counts. Essentially, this test is utilized to assess the association between types of feet and clustered groups of goalkeepers.

1.5 Cluster Analysis

1.5.1 Hierarchical Agglomerative Cluster Analysis (HACA)

Hierarchical agglomerative cluster analysis serves both as an exploratory tool and a non-exploratory technique. It establishes a cluster hierarchy for a single observation and groups related observations into distinct observations [37]. It is worth noting that in this algorithm, the learning process is driven by both dataset merges and splits [38–40]. These operations help highlight similar findings in a dendrogram. It is important to emphasize that the number of clusters indicated by HACA depends on the proximity of a specific or predetermined cluster within the dendrogram. In this analysis, the cosine distance was utilized, and the clustering validation technique was carried out using class centroids [41].

1.5.2 Louvain Clustering

The most recent clustering algorithm for categorizing a provided dataset or observations is the Louvain clustering technique. This method is structured to accomplish its task through two distinct phases. Initially, it aims to identify a "sparse" group by optimizing modularity using a traditional approach. In the subsequent phase, the algorithm links nodes from interconnected communities to establish a unique community, leading to the creation of a fresh network of community nodes [42]. These steps can be repeated repeatedly until a modularity condition is met. This step also adds to the system's hierarchical fragmentation and the production of many divisions [43]. The divisions are often based on the density of the communities' borders, rather than the intercommunity margins.

1.5.3 K-Means Cluster Analysis

The k-means clustering technique is a type of cluster analysis methodology that segregates a dataset into k predetermined and distinct subgroups known as clusters, assigning each data point to a single group [39, 44, 45]. The objective of the method is to establish strong connections among intercluster data points while ensuring that intracluster data points remain distinct. Cluster analysis was employed to categorize the data into groups, using performance indicators evaluated in this study. It is important to highlight that the Euclidean distance was utilized as a distance metric to determine the composition of all clusters formed in the study.

1.6 Principal Component Analysis as Data Mining Technique

Principal component analysis (PCA) is a mathematical technique primarily used to uncover underlying patterns within a dataset comprising observed variables [39, 44]. PCA facilitates the identification of significant variables that potentially capture the essence of a given dataset, by examining the spatial and temporal variations present in the data. The process of extracting insights through PCA involves eliminating data associated with the least influential component and retaining the most valuable information within the dataset [46, 47]. The application of PCA becomes essential in efficiently reducing the dimensionality of a large dataset, a critical aspect as it aids in preventing wastage of resources, expenses, and time, given that the original data is often preserved.

1.7 Application of Machine Learning Models in the Study

In this brief, diverse supervised machine learning models were employed to classify various investigated classes. These models include k-nearest neighbours (kNN), support vector machine (SVM), logistic regression (LR), decision tree (DT), and artificial neural networks (ANN). It is important to emphasize that, at times, the hyperparameters of these evaluated models were optimized, and such occurrences are explicitly indicated in the subsequent chapters where machine learning models are employed. Readers are encouraged to refer to our previous works [1, 48, 49]. For detailed explanations of the aforementioned machine learning models, as well as the performance metrics frequently used to assess the effectiveness of the developed machine learning pipeline.

1.8 Datasets for the Study

A dataset of 1601 goalkeepers from the top five European leagues (English Premier League, Spanish Laliga, Italian Serie A, French Ligue1, and German Bundesliga) across five consecutive seasons was utilized. The dataset included information on clean sheets and anthropometric variables such as weight, height, BMI, and the dominant foot of the goalkeepers (right, left or both). Before the full analysis in this study, the data underwent pre-processing and thorough checking for any missing information. Any rows that contained missing information were removed from the date as recommended by the preceding investigators [50, 51].

Due to the relative importance of goalkeepers' statistics for both the national federation and teams, InStat developed and covered over 63 indicators for goalkeepers and to ensure the reliability and validity of the indicators, and each indicator is linked to videos that provide the activity profile of the goalkeepers as well as comprehensive reports on the goalkeepers' overall activity in the match [52]. This information is vital as it portrays the extent to which goalkeepers influence a match. Table 1.1 depicts a detailed description of the metrics used in the study.

Table 1.1 Full metrics used in the study

Main statistics	Technical and tactical skills	Opponents statistics
Matches played	Passes	Close-range shots
Minutes played	Accurate passes	Close-range shots on target
Clean sheets	Accurate passes %	Close-range shots saved
Opponents' shots	Key passes	Close-range shots saved, %
Opponents' shot on target	Foot passes from open play	Mid-range shots
Goals conceded	Accurate passes from open play	Mid-range shots on target
Shots saved	Hand passes accurate	Mid-range shots saved
Shots saved, %	Hand passes accurate %	Mid-range shots saved, %
Super saves	Passes from set pieces	Long-range shots
Penalties saved	Accurate passes from set pieces	Long-range shots on target
Goals conceded—penalties	Set piece passes accurate, %	Long-range shots saved
Penalties saved, %	Short passes	% of long-range shots saved
Goalkeepers interception	Short passes accurate	Jumping saves
Good interception of GKs	Short passes accurate, %	Saves without jumping
Goalkeepers' interception	Medium passes	Stopped shots

(continued)

Table 1.1 (continued)

Main statistics	Technical and tactical skills	Opponents statistics
Goalkeepers' interception %	Medium passes accurate	Stopped shots %
	Mid-range passes accurate	Shots saves with successful bouncing
Anthropometric	Long passes	Shots saves with unsuccessful bouncing
Age	Accurate long passes	
Height	Accurate long passes	
Weight	Accurate long passes, %	
Foot		

References

1. R.M. Musa, A.P.P.A. N.A. Majeed, Kosni, M.R. Abdullah, Machine Learning in Team Sports: Performance Analysis and Talent Identification in Beach Soccer & Sepak-takraw. Springer Nature (2020)
2. B. Noël, J. Van der Kamp, S. Klatt, The interplay of goalkeepers and penalty takers affects their chances of success. Front. Psychol. **12**, 645312 (2021)
3. G. Gelade, Evaluating the ability of goalkeepers in English Premier League football. J. Quant. Anal. Sport. **10**, 279–286 (2014)
4. M.R. Abdullah, A. Maliki, R.M. Musa, N.A. Kosni, H. Juahir, Intelligent prediction of soccer technical skill on youth soccer player's relative performance using multivariate analysis and artificial neural network techniques. Int. J. Adv. Sci. Eng. Inf. Technol. **6**, 668–674 (2016)
5. M.R. Abdullah, R.M. Musa, A. Maliki, P.K. Suppiah, N.A. Kosni, Relationship of physical characteristics, mastery and readiness to perform with position of elite soccer players. Int. J. Adv. Eng. Appl. Sci. **1**, 8–11 (2016)
6. R.M. Musa, A.P.P. Abdul Majeed, M.R. A.F.A.B. Abdullah, Nasir, M.H.A. Hassan, M.A.M. Razman, Technical and tactical performance indicators discriminating winning and losing team in elite Asian beach soccer tournament. PLoS One. 14 (2019). https://doi.org/10.1371/journal.pone.0219138
7. R. Muazu Musa, A.P.P. Abdul Majeed, M.R. Abdullah, G. Kuan, M.A. Mohd Razman, Current trend of analysis in high-performance sport and the recent updates in data mining and machine learning application in sports, in *Data Mining and Machine Learning in High-Performance Sport* (Springer, 2022), pp. 1–11
8. H. Sarmento, R. Marcelino, M.T. Anguera, J. Campaniço, N. Matos, J.C. LeitÃo, Match analysis in football: a systematic review. J. Sports Sci. **32**, 1831–1843 (2014)
9. M.A. Abdullah, M.A.R. Ibrahim, M.N.A Bin Shapiee, M.A. Mohd Razman, R.M. Musa, A.P.P. Abdul Majeed, The classification of skateboarding trick manoeuvres through the integration of IMU and machine learning, in Lecture Notes in Mechanical Engineering. pp. 67–74 (2020). https://doi.org/10.1007/978-981-13-9539-0_7
10. Z. Taha, R.M. Musa, A.P.P. Abdul Majeed, M.R. Abdullah, M.A. Zakaria, M.M. Alim, J.A.M. Jizat, M.F. Ibrahim, The identification of high potential archers based on relative psychological coping skills variables: a support vector machine approach, in *IOP Conference Series: Materials Science and Engineering* (2018). https://doi.org/10.1088/1757-899X/319/1/012027
11. J. Santos, P.M. Sousa, V. Pinheiro, F.J. Santos, Analysis of offensive and defensive actions of young soccer goalkeepers. Hum. Mov. **23**, 18–27 (2022)
12. J.A. Navia Manzano, L.M. Ruiz Perez, On the use of situational and body information in goalkeeper actions during a soccer penalty kick. Int. J. Sport Psychol. **44**, 234–251 (2013)

13. S.M. Gil, J. Zabala-Lili, I. Bidaurrazaga-Letona, B. Aduna, J.A. Lekue, J. Santos-Concejero, C. Granados, Talent identification and selection process of outfield players and goalkeepers in a professional soccer club. J. Sports Sci. **32**, 1931–1939 (2014)
14. F. Hammes, A. Hagg, A. Asteroth, D. Link, Artificial intelligence in elite sports—a narrative review of success stories and challenges. Front. Sport. Act. Living. **4** (2022)
15. A. Rossi, L. Pappalardo, P. Cintia, A narrative review for a machine learning application in sports: an example based on injury forecasting in soccer. Sports **10**, 5 (2021)
16. H. Van Eetvelde, L.D. Mendonça, C. Ley, R. Seil, T. Tischer, Machine learning methods in sport injury prediction and prevention: a systematic review. J EXP ORTOP. **8**, 27 (2021). https://doi.org/10.1186/s40634-021-00346-x
17. R. Muazu Musa, A.P.P. Abdul Majeed, M.Z. Suhaimi, M.A. Mohd Razman, M.R. Abdullah, N.A. Abu Osman, Nature of volleyball sport, performance analysis in volleyball, and the recent advances of machine learning application in sports. Mach. Learn. Elit. Volleyb. 1–11 (2021)
18. M. Herold, F. Goes, S. Nopp, P. Bauer, C. Thompson, T. Meyer, Machine learning in men's professional football: current applications and future directions for improving attacking play. Int. J. Sports Sci. Coach. **14**, 798–817 (2019)
19. J.G. Claudino, D. de O. Capanema, T.V. de Souza, J.C. Serrão, A.C. Machado Pereira, G.P. Nassis, Current approaches to the use of artificial intelligence for injury risk assessment and performance prediction in team sports: a systematic review. Sport. Med. **5**, 1–12 (2019)
20. C. Richter, M. O'Reilly, E. Delahunt, Machine learning in sports science: challenges and opportunities. Sport. Biomech. 1–7 (2021). https://doi.org/10.1080/14763141.2021.1910334
21. F.Y. Tan, M.H.A. Hassan, A.P.P. Abdul Majeed, M.A. Mohd Razman, M.A. Abdullah, Classification of Sepak Takraw kicks using machine learning, in *Human-Centered Technology for a Better Tomorrow* (Springer, 2022), pp. 321–331
22. F. Thabtah, L. Zhang, N. Abdelhamid, NBA game result prediction using feature analysis and machine learning. Ann. Data Sci. **6**, 103–116 (2019)
23. J.M. Echterhoff, J. Haladjian, B. Brugge, Gait and jump classification in modern equestrian sports, in *Proceedings—International Symposium on Wearable Computers, ISWC.* pp. 88–91 (2018). https://doi.org/10.1145/3267242.3267267
24. J.W. McGrath, J. Neville, T. Stewart, J. Cronin, Cricket fast bowling detection in a training setting using an inertial measurement unit and machine learning. J. Sports Sci. **37**, 1220–1226 (2019)
25. M.T.O. Worsey, H.G. Espinosa, J.B. Shepherd, D.V Thiel, An evaluation of wearable inertial sensor configuration and supervised machine learning models for automatic punch classification in boxing. IoT. **1**, 360–381 (2020)
26. M.S.M. Duki, M.N.A. Shapiee, M.A. Abdullah, I.M. Khairuddin, M.A.M. Razman, A.P.P.A. Majeed, The classification of taekwondo kicks via machine learning: a feature selection investigation. MEKATRONIKA **3**, 61–67 (2021)
27. M.A. Abdullah, M.A.R. Ibrahim, M.N.A. Shapiee, M.A. Zakaria, M.A.M. Razman, R.M. Musa, N.A.A. Osman, A.P.P.A. Majeed, The classification of skateboarding tricks via transfer learning pipelines. PeerJ Comput. Sci. **7**, e680 (2021)
28. R.M. Musa, A.P.P. Abdul Majeed, M.Z. Suhaimi, M.R. Abdullah, M.A. Mohd Razman, D. Abdelhakim, N.A. Abu Osman, Identification of high-performance volleyball players from anthropometric variables and psychological readiness: a machine-learning approach. Proc. Inst. Mech. Eng. Part P J. Sport. Eng. Technol (2021). https://doi.org/10.1177/175433712110 45451
29. R. Muazu Musa, A.P.P. Abdul Majeed, M.R. Abdullah, A.F. Ab. Nasir, M.H. Arif Hassan, M.A. Mohd Razman, Technical and tactical performance indicators discriminating winning and losing team in elite Asian beach soccer tournament. PLoS One. **14**, e0219138 (2019)
30. T.W. MacFarland, J.M. Yates, Mann–whitney u test, in, *Introduction to Nonparametric Statistics for the Biological Sciences Using R* (Springer, 2016), pp. 103–132
31. R.M. Musa, A.P.P.A. Majeed, N.A. Kosni, M.R. Abdullah, Technical and tactical performance indicators determining successful and unsuccessful team in elite beach soccer, in *Machine Learning in Team Sports* (Springer, 2020), pp. 21–28

32. D. Taylor, T. Williams, Sports injuries in athletes with disabilities: wheelchair racing. Spinal Cord. **33**, 296–299 (1995)
33. P.K. Suppiah, J.L.F. Lee, A.M.N. Azmi, H. Noordin, R.M. Musa, Relative age effect in U-16 Asian championship soccer tournament. Malaysian J. Movement, Heal. Exerc. **9** (2020)
34. R.M. Musa, I. Hassan, M.R. Abdullah, M.N.L. Azmi, A.P.P.A. Majeed, N.A.A. Osman, A longitudinal analysis of injury characteristics among elite and amateur tennis players at different tournaments from electronic newspaper reports. Front. Public Heal. **10** (2022)
35. R.M. Musa, I. Hassan, M.R. Abdullah, M.N.L. Azmi, A.P.P. Abdul Majeed, N.A. Abu Osman, Surveillance of injury types, locations, and intensities in male and female tennis players: a content analysis of online newspaper reports. Int. J. Environ. Res. Public Health. **18**, 12686 (2021)
36. I. Hassan, R.M. Musa, M.N. Latiff Azmi, M. Razali Abdullah, S.Z. Yusoff, Analysis of climate change disinformation across types, agents and media platforms. Inf. Dev. 02666669221148693 (2023)
37. O. Maimon, L. Rokach, Data Mining and Knowledge Discovery Handbook. (2005). https://doi.org/10.1007/b107408
38. A.B.H.M. Maliki, M.R. Abdullah, H. Juahir, F. Abdullah, N.A.S. Abdullah, R.M. Musa, S.M. Mat-Rasid, A. Adnan, N.A. Kosni, W.S.A.W. Muhamad, N.A.M. Nasir, A multilateral modelling of Youth Soccer Performance Index (YSPI). IOP Conf. Ser. Mater. Sci. Eng. **342**, 012057 (2018). https://doi.org/10.1088/1757-899X/342/1/012057
39. M.R. Razali, N. Alias, A. Maliki, R.M. Musa, L.A. Kosni, H. Juahir, Unsupervised pattern recognition of physical fitness related performance parameters among Terengganu youth female field hockey players. Int. J. Adv. Sci. Eng. Inf. Technol. **7**, 100–105 (2017)
40. Z. Taha, R.M. Musa, M.R. Abdullah, A. Maliki, N.A. Kosni, S.M. Mat-Rasid, A. Adnan, H. Juahir, Supervised pattern recognition of archers' relative psychological coping skills as a component for a better archery performance. J. Fundam. Appl. Sci. **10**, 467–484 (2018)
41. R. Muazu Musa, A.P.P. Abdul Majeed, Z. Taha, M.R. Abdullah, A.B. Husin Musawi Maliki, N. Azura Kosni, The application of artificial neural network and k-nearest neighbour classification models in the scouting of high-performance archers from a selected fitness and motor skill performance parameters. Sci. Sport. (2019). https://doi.org/10.1016/j.scispo.2019.02.006
42. C. Wu, R.C. Gudivada, B.J. Aronow, A.G. Jegga, Computational drug repositioning through heterogeneous network clustering. BMC Syst. Biol. **7**, S6 (2013). https://doi.org/10.1186/1752-0509-7-S5-S6
43. V.D. Blondel, J. Guillaume, R. Lambiotte, E. Lefebvre: fast unfolding of community hierarchies in large networks. J. Stat. Mech. theory Exp. 10008 (2008). https://doi.org/10.1088/1742-5468/2008/10/P10008
44. V. Eswaramoorthi, M.R. Abdullah, R.M. Musa, A.B.H.M. Maliki, N.A. Kosni, N.B. Raj, N. Alias, H. Azahari, S.M. Mat-Rashid, H. Juahir, A multivariate analysis of cardiopulmonary parameters in archery performance. Hum. Mov. **19**, 35–41 (2018). https://doi.org/10.5114/hm.2018.77322
45. Z. Taha, M. Haque, R.M. Musa, M.R. Abdullah, A. Maliki, N. Alias, N.A. Kosni, Intelligent prediction of suitable physical characteristics toward archery performance using multivariate techniques. J Glob Pharma Technol. **9**, 44–52 (2009)
46. R.M. Musa, M.R. Abdullah, A.B.H.M. Maliki, N.A. Kosni, M. Haque, The application of principal components analysis to recognize essential physical fitness components among youth development archers of Terengganu, Malaysia. Indian J. Sci. Technol. **9** (2016)
47. R.M. Musa, M.R. Abdullah, A.B.H.M. Maliki, N.A. Kosni, S.M. Mat-Rasid, A. Adnan, H. Juahir, Supervised pattern recognition of archers' relative psychological coping skills as a component for a better archery performance. J. Fundam. Appl. Sci. **10**, 467–484 (2018)
48. R.M. Musa, A.P.P.A. Majeed, M.Z. Suhaimi, M.A.M. Razman, M.R. Abdullah, N.A.A. Osman, *Machine learning in elite volleyball: integrating performance analysis* (Springer, Competition and Training Strategies, 2021)
49. R. Muazu Musa, Z. Taha, A.P.P. Abdul Majeed, M.R. Abdullah, Machine Learning in Sports (2019). https://doi.org/10.1007/978-981-13-2592-2

50. M.R. Abdullah, V. Eswaramoorthi, R.M. Musa, A.B.H.M. Maliki, N.A. Kosni, M. Haque, The effectiveness of aerobic exercises at difference intensities of managing blood pressure in essential hypertensive information technology officers. J. Young Pharm. **8**, 483–486 (2016). https://doi.org/10.5530/jyp.2016.4.27

51. M.A. Gipit, M.R.A. Charles, R.M. Musa, N.A. Kosni, A.B.H.M. Maliki, The effectiveness of traditional games intervention programme in the improvement of form one school-age children's motor skills related performance components. Movement, Heal. Exerc. **6**, 157–169 (2017)

52. INSTAT FOR REFEREES—InStat, https://instatsport.com/football/for_referees. Last accessed 11 Aug 2022

Chapter 2
The Influence of Anthropometrics on Goalkeepers' Penalty-Saving Performance in Elite European Football

Abstract In this chapter, we examined the important anthropometric parameters associated with goalkeepers' penalty-saving performance in European championship tournaments. It was demonstrated from the findings of the study that certain anthropometric variables including weight, height, BMI, and age are discriminating factors for a successful penalty-saving performance. The high penalty savers are essentially taller, older, and heavier in comparison with the low penalty savers. In addition, it was demonstrated that the use of feet could play a role in penalty-saving performance. The high penalty savers commonly used their right legs as a dominant foot, while larger proportions of the low penalty-saving GKs are left-footed and a few used both feet.

Keywords Anthropometric parameters · Goalkeepers' performance · Penalty shootouts · European football championships

2.1 Overview

In modern football, especially at the elite level such as European football, the role of goalkeepers (GKs) is of utmost importance in influencing the outcome of closely contested matches [1]. Their proficiency in saving penalty kicks, which are considered one of the most intense and pivotal moments on the pitch, can significantly alter the fortunes of their respective teams. In recent years, the evolution of football has witnessed a greater emphasis on data-driven analyses to comprehend the factors that contribute to a goalkeeper's success in saving penalties [2]. Among such factors, anthropometric attributes have emerged as critical determinants of a goalkeeper's penalty-saving prowess.

Anthropometrics, which include measurements such as height, arm span, and body mass index (BMI), have long been a subject of discussion for researchers and coaches alike [3, 4]. On the other hand, some studies focused on the physical characteristics and fitness profiles of GKs in conjunction with on-field players. For instance, Rebelo et al. [5] reported that there were some differences in anthropometric characteristics and physical fitness between different positions of players. GKs were taller and heavier than players in other positions, while midfielders were the least tall and heavy. Central defenders had the highest body fat percentage, while forwards had the lowest. Other researchers focused on the analysis of the offensive and defensive (tactical–technical) actions of GKs [6, 7].

Saving a penalty kick is an important action that could be described as an interceptive ability to stop a ball from entering the net during a penalty shootout. The relationship between physical attributes and a GK's penalty-saving performance has sparked a passionate debate within the football community. While some argue that taller goalkeepers possess a greater reach, thereby enhancing their ability to prevent penalty shots, others maintain that factors such as agility and reaction time may be equally significant regardless of height.

The current investigation aims to elucidate the correlation between anthropometrics and penalty-saving performance among elite European football goalkeepers. The study is geared towards providing insights on the link between the variables that could aid in the strategic selection and training of GKs thereby ultimately enhancing their proficiency in upsetting penalty attempts from the opponents.

2.2   Data Treatment

We utilized a dataset comprising 1599 GKs from the top five European leagues, namely the English Premier League, Spanish Laliga, Italian Serie A, French Ligue 1, and German Bundesliga, across five consecutive seasons. These datasets contained information on penalty saving and anthropometric variables such as weight, height, BMI, and the dominant foot of the goalkeepers, i.e. right, left, or both. It is worth highlighting that the data underwent pre-processing and thorough checking for any missing information before the commencement of the full analysis in this study. Any rows that contained any missing information were removed from the dataset [8, 9].

2.3   Clustering

The Louvain clustering method is a modern algorithm that can group data or observations into categories. It works in two steps: first, it finds small clusters by maximizing modularity in a standard way. Second, it merges nodes that belong to the same cluster into a new node, creating a new network of cluster nodes [10–12]. In the current study, Louvain analysis was used to group the performance of the GKs in saving the penalties based on occurrences, resulting in two clusters: high penalty savers (HPS) and low penalty savers (LPS). The Euclidean distance was used in separating the groups. The Mann–Whitney U-test was then applied to examine the variation of the GKs group with respect to the anthropometric variables and penalty-saving performances.

2.4   Model Development

In the present study, there exists an imbalance in the existing dataset, i.e. there are 743 LPS and 858 HPS. Therefore, the dataset was under-sampled owing to the sizeable dataset having an even distribution for both the LPS and HPS with 743 for each class. The remaining dataset was split into the 80:20 ratio for training and testing, respectively [13–15]. Two machine learning models were developed and evaluated for their efficacy in demarcating the classes, namely logistic regression (LR) and decision tree (DT). The models were evaluated based on their classification accuracy as well as the confusion matrix. Spyder IDE v 5.4.3 running on Python 3.9 was used to evaluate the models. Default hyperparameters of the models were used in the present study.

2.5   Results and Discussion

The anthropometric and penalty-saving performance variations between the two groups of GKs are given in Table 2.1. It could be seen from the table that HPS was older with mean and medium age of 34 in comparison with the LPS with 29. The HPS are also found to be heavier and taller than the LPS p (< 0.001).

Figure 2.1 depicts the results of the chi-square analysis on the differences between the GKs based on the dominant foot used. Statistically significant differences were observed between the two groups of GKs in which the HPS are found to be dominantly right-footed, a certain number of the GKs used left, while a little proportion used both feet χ^2 (2) $= 19.7$, $p < 0.001$. This finding demonstrates that high penalty-saving GKs are mostly right-dominant.

Figure 2.2 illustrates the efficacy of the models developed, i.e. LR and DT. It is apparent from the figure that the DT has a better capability in discerning the LPS and HPS classes with a training and testing accuracy of 95% and 93%, respectively. The confusion matrix depicted in Fig. 2.3 provides further insight into the ability DT model to classify the two groups of the GKs.

Table 2.1 Anthropometric and performance differences of the goalkeepers' group

Anthropometric	Group	N	Mean	Median	SD	p-value
Age	HPS	858	33.8	34	4.78	0.001*
	LPS	743	29.3	29	4.63	
Weight	HPS	858	86.7	87	5.03	0.001*
	LPS	743	77.4	80	14.64	
Height	HPS	858	191.6	192	4.05	0.001*
	LPS	743	187.3	188	7.85	

* Mann–Whitney U-test, $p < 0.001$

Fig. 2.1 Dominant foot and goalkeepers' penalty-saving performance

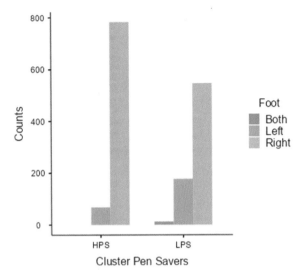

Fig. 2.2 Testing and training accuracy of the developed models

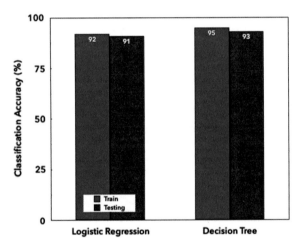

The findings from the present investigation revealed that high penalty-saving GKs in the top European soccer leagues are attributed to higher anthropometry and older. Essentially, the GKs are heavier, taller, and comparatively older than the low-saving counterparts. Several studies have shown that elite goalkeepers are taller and heavier than non-elite goalkeepers [4, 16]. It is worth noting that the function of athletes on the playing field, such as those possessing speed or power, necessitates that they adapt their traits in accordance with the requirements of their designated position [8, 17–19]. The increased age of elite goalkeepers is likely due to the fact that they have had more time to develop their skills and experience. They are also more likely to be physically mature, which gives them an advantage in terms of strength and power. Therefore, the anthropometric characteristics of elite goalkeepers, such as height,

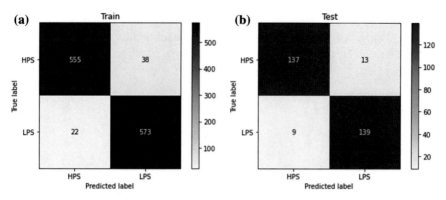

Fig. 2.3 Train **a** and test **b** confusion matrix of the DT model

weight, and age, are related to their penalty-saving ability. These characteristics give goalkeepers an advantage in terms of reach, strength, and power, which can help them to save penalties.

The data has further indicated that the most successful penalty savers are predominantly right-footed. It could be speculated that right-footed goalkeepers possess an innate advantage when it comes to preventing penalties taken by right-footed shooters. Due to their shared dominant side, right-footed goalkeepers could potentially have an easier time predicting the shot's direction and trajectory, which leads to a greater success rate in halting penalties from right-footed players. Additionally, penalty-taking involves both a physical and psychological battle between the shooter and the goalkeeper. Right-footed goalkeepers may psychologically intimidate right-footed shooters, as they are aware that they are up against an opponent with a similar dominant foot [20]. This psychological edge could cause the shooter to experience heightened pressure, resulting in less precise and weaker penalties. Moreover, the machine learning models as well as the DT model performed excellently well in predicting the two groups of the GKs. The DT was shown to be an effective algorithm for both classification and regression analysis [12].

2.6 Summary

The present investigation has identified the essential anthropometric variables contributing to GK penalty-saving performance. Certain anthropometric variables including weight, height, BMI, and age are found to be discriminating factors for a successful penalty-saving performance. The high penalty savers are essentially taller, older, and heavier in comparison with the low penalty savers. In addition, it was demonstrated that the use of feet could play a role in penalty-saving performance. The high penalty savers commonly used their right legs as a dominant foot, while larger proportions of the low penalty-saving GKs are left-footed, and a few used both feet. The findings from the current study could provide insight into the vital anthropometric variables contributing to a successful penalty-saving performance that could aid in the strategic selection and training of goalkeepers that may assist in enhancing GKs' proficiency in upsetting penalty attempts from the opponents.

References

1. M. Dicks, C. Button, K. Davids, Examination of gaze behaviors under in situ and video simulation task constraints reveals differences in information pickup for perception and action. Attention, Perception, Psychophys. **72**, 706–720 (2010)
2. C.H. Almeida, A. Volossovitch, R. Duarte, Penalty kick outcomes in UEFA club competitions (2010–2015): the roles of situational, individual and performance factors. Int. J. Perform. Anal. Sport **16**, 508–522 (2016)
3. M.R. Abdullah, R.M. Musa, A. Maliki, P.K. Suppiah, N.A. Kosni, Relationship of physical characteristics, mastery and readiness to perform with position of elite soccer players. Int. J. Adv. Eng. Appl. Sci. **1**, 8–11 (2016)
4. M.R. Abdullah, A. Maliki, R.M. Musa, N.A. Kosni, H. Juahir, Intelligent prediction of soccer technical skill on youth soccer player's relative performance using multivariate analysis and artificial neural network techniques. Int. J. Adv. Sci. Eng. Inf. Technol. **6**, 668–674 (2016)
5. A. Rebelo, J. Brito, J. Maia, M.J. Coelho-e-Silva, A.J. Figueiredo, J. Bangsbo, R.M. Malina, A. Seabra, Anthropometric characteristics, physical fitness and technical performance of under-19 soccer players by competitive level and field position. Int. J. Sports Med. 312–317 (2012)
6. F. Santos, J. Santos, M. Espada, C. Ferreira, P. Sousa, V. Pinheiro, T-pattern analysis of offensive and defensive actions of youth football goalkeepers. Front. Psychol. 5371 (2022)
7. J. Santos, P.M. Sousa, V. Pinheiro, F.J. Santos, Analysis of offensive and defensive actions of young soccer goalkeepers. Hum. Mov. **23**, 18–27 (2022)
8. M.R. Abdullah, V. Eswaramoorthi, R.M. Musa, A.B.H.M. Maliki, N.A. Kosni, M. Haque, The effectiveness of aerobic exercises at difference intensities of managing blood pressure in essential hypertensive information technology officers. J. Young Pharm. **8**, 483–486 (2016). https://doi.org/10.5530/jyp.2016.4.27
9. M.A. Gipit, M.R.A. Charles, R.M. Musa, N.A. Kosni, A.B.H.M. Maliki, The effectiveness of traditional games intervention programme in the improvement of form one school-age children's motor skills related performance components. Movement, Heal. Exerc. **6**, 157–169 (2017)
10. R. Muazu Musa, Z. Taha, A.P.P. Abdul Majeed, M.R. Abdullah, Psychological variables in ascertaining potential archers, in *SpringerBriefs in Applied Sciences and Technology* (2019), pp. 21–27. https://doi.org/10.1007/978-981-13-2592-2_3
11. R.M. Musa, A.P.P.A. Majeed, N.A. Kosni, M.R. Abdullah, Machine learning in team sports: performance analysis and talent identification in beach soccer & Sepak-takraw (Springer Nature, 2020)
12. R. Muazu Musa, A.P.P. Abdul Majeed, M.R. Abdullah, G. Kuan, M.A. Mohd Razman, Current trend of analysis in high-performance sport and the recent updates in data mining and machine learning application in sports, in *Data Mining and Machine Learning in High-Performance Sport* (Springer, 2022), pp. 1–11
13. Z. Taha, R.M. Musa, A.P.P. Abdul Majeed, M.R. Abdullah, M.A. Zakaria, M.M. Alim, J.A.M. Jizat, M.F. Ibrahim, The identification of high potential archers based on relative psychological coping skills variables: a support vector machine approach, in *IOP Conference Series: Materials Science and Engineering* (2018). https://doi.org/10.1088/1757-899X/319/1/012027
14. M.A. Abdullah, M.A.R. Ibrahim, M.N.A. Bin Shapiee, M.A. Mohd Razman, R.M. Musa, A.P.P. Abdul Majeed, The classification of skateboarding trick manoeuvres through the integration

of IMU and machine learning, in Lecture Notes in Mechanical Engineering (2020), pp. 67–74. https://doi.org/10.1007/978-981-13-9539-0_7

15. Z. Taha, M.A.M. Razman, F.A. Adnan, A.S. Abdul Ghani, A.P.P. Abdul Majeed, R.M. Musa, M.F. Sallehudin, Y. Mukai, The identification of hunger behaviour of Lates Calcarifer through the integration of image processing technique and support vector machine, in *IOP Conference Series: Materials Science and Engineering* (2018). https://doi.org/10.1088/1757-899X/319/1/012028

16. S.M. Gil, J. Zabala-Lili, I. Bidaurrazaga-Letona, B. Aduna, J.A. Lekue, J. Santos-Concejero, C. Granados, Talent identification and selection process of outfield players and goalkeepers in a professional soccer club. J. Sports Sci. **32**, 1931–1939 (2014)

17. V. Eswaramoorthi, M.R. Abdullah, R.M. Musa, A.B.H.M. Maliki, N.A. Kosni, N.B. Raj, N. Alias, H. Azahari, S.M. Mat-Rashid, H. Juahir, A multivariate analysis of cardiopulmonary parameters in archery performance. Hum. Mov. **19**, 35–41 (2018). https://doi.org/10.5114/hm.2018.77322

18. M.A. Gipit Charles, M.R. Abdullah, R.M. Musa, N.A. Kosni, A.B.H.M. Maliki, The effectiveness of traditional games intervention program in the improvement of form one school-age children's motor skills related performance components. J. Phys. Educ. Sport. **17**, 925–930 (2017). https://doi.org/10.7752/jpes.2017.s3141

19. R.M. Musa, A.P.P. Abdul Majeed, M.R. Abdullah, A.F.A.B. Nasir, M.H.A. Hassan, M.A.M. Razman, Technical and tactical performance indicators discriminating winning and losing team in elite Asian beach soccer tournament. PLoS One. **14** (2019). https://doi.org/10.1371/journal.pone.0219138

20. J.A. Navia Manzano, L.M. Ruiz Perez, On the use of situational and body information in goalkeeper actions during a soccer penalty kick. Int. J. Sport Psychol. **44**, 234–251 (2013)

Chapter 3
The Relationship Between Anthropometrics Parameters and Clean Sheets of Goalkeepers in Elite European Football

Abstract This chapter highlights the association between goalkeepers' (GKs) clean sheet performances and their anthropometric indexes. It has been demonstrated from the study findings that GKs with bigger body sizes and older ages are more likely to keep clean sheets compared to those with lower anthropometrics and younger ages. The study also found that the feet used by the GKs have no association with keeping clean sheets. Therefore, higher anthropometric indexes and clean sheets in goalkeepers highlight the multidimensional nature of goalkeeping performance. It suggests that a combination of physical attributes as well as experience can contribute to a goalkeeper's ability to keep clean sheets. It has also been demonstrated from the current finding that the logistic regression model is able to yield an excellent prediction of the group of GKs based on the investigated parameters. Understanding these factors could improve defensive performance in football and assist stakeholders in making better decisions about who to recruit and develop as GKs.

Keywords Football goalkeepers · Clean sheets · Anthropometric parameters · Dominant foot · Logistic regression model · European football league

3.1 Overview

In elite European football, the role of goalkeepers (GKs) is of utmost importance in determining the success of their respective teams. The ability of goalkeepers to maintain a clean sheet, i.e. to prevent the opposing team from scoring, is often regarded as a benchmark of their skill and efficiency [1]. It is essential to recognize that not all clean sheets hold the same level of impressiveness, as explained previously [1]. Thus, a thorough assessment of a goalkeeper's performance requires a more nuanced metric. It is worth noting that statistical data regarding GKs is relatively limited when compared to that of outfield players [2]. Nevertheless, some factions in the football community still view clean sheets as the primary indicator of a goalkeeper's performance.

Research in the field of sports science has shown that anthropometric parameters can have a profound impact on athletic performance in various sports disciplines [3–6]. For instance, studies have demonstrated that taller basketball players have an advantage in rebounding and shot-blocking due to their increased reach and ability to contest shots [7]. Similarly, in football, height has been associated with improved aerial ability and commanding presence in the penalty area [5, 8].

Despite extensive research on factors such as technique, agility, and positioning concerning GK performance, the impact of anthropometric parameters on clean sheets has remained relatively unexplored. Anthropometric parameters, such as height, weight, body mass index, and limb length, are inherent physical attributes that can significantly affect an individual's athletic performance. A comprehension of the relationship between these parameters and clean sheets in elite European football can provide valuable insights for player selection, training, and performance optimization. However, the specific influence of anthropometric parameters on the clean sheet record of goalkeepers in elite European football remains largely unexplored. The purpose of this study is to investigate the relationship between anthropometric parameters and clean sheets of GKs in elite European football. The findings of this study will provide valuable information for coaches and goalkeepers on how to improve their performance and keep more clean sheets.

To achieve the purpose of the study, we analysed a dataset of 1601 goalkeepers from the top five European leagues (English Premier League, Spanish Laliga, Italian Serie A, French Ligue1, and German Bundesliga) across five consecutive seasons. The dataset included information on clean sheets and anthropometric variables such as weight, height, BMI, and the dominant foot of the goalkeepers (right, left, or both). Before the full analysis of this study began, the data underwent pre-processing and thorough checking for any missing information. Any rows that contained missing information were removed from the dataset [9, 10].

3.2 Clustering

The Louvain clustering method is a modern algorithm that can group data or observations into categories. It works in two steps: first, it finds small clusters by maximizing modularity in a standard way. Second, it merges nodes that belong to the same cluster into a new node, creating a new network of cluster nodes [11–13]. In this study, we used the Louvain cluster analysis to group the GKs' clean sheets based on how often they occurred during matches. The Euclidean distance was applied as a measure to determine the formation of two clusters: high and low clean sheets (HCS, LCS) as a target for the final output of the clusters.

3.3   Machine Learning-Based Classification Model

In this investigation, there is a class imbalance, where from the 1601 samples, 730 fall into the LCS class, while the remaining 871 are in the HCS class. Owing to the size of the dataset, an oversampling approach is taken to mitigate the imbalance, i.e. the Synthetic Minority Oversampling TEchnique (SMOTE) was applied to have an even distribution of 871 for both classes. The new dataset was split into the 80:20 ratio for training and testing, respectively [14–16]. Three machine learning classification algorithms, viz., logistic regression (LR), decision tree (DT), and support vector machine (SVM) were implemented and assessed for their effectiveness in discriminating between classes. The evaluation metrics utilized were overall classification accuracy as well as confusion matrix analysis. The programming environment consisted of Spyder IDE version 5.4.3 integrated with Python 3.9. The default hyperparameter (vanilla) configurations were employed for each of the models, i.e. without tuning. The comparative analysis focused on the predictive performance of unseen test data across the methods examined. Both classifier precision and differential success across classes were examined via the confusion matrices generated.

3.4   Results and Discussion

Figure 3.1 depicts the classes identified through the k-means analysis with respect to the clean sheets recorded by the two groups of the GKs. It could be seen from the figure that a clear partition was established between the HCS and LCS. In other words, the mean occurrences of the clean sheets accrued by the GKs are substantially higher in HCS as compared to the LCS.

The anthropometric variations between the two groups of GKs are displayed in Fig. 3.2. It could be seen from the table that the HCS was older and also found to be heavier and taller than the LCS. This finding reflects that keeping clean sheets is associated with high levels of anthropometry and maturity.

Figure 3.3 depicts the results of the chi-square analysis on the differences between the GKs based on the dominant foot used. No statistically significant differences were observed between the two groups of GKs $\chi^2 (2) = 1.31$, $p = 0.518$. This finding demonstrates that keeping clean sheets is not associated with the type of feet used. In other words, GKs could keep a clean sheet record irrespective of their dominant feet.

Figure 3.4 shows the classification performance of the three models, namely logistic regression (LR), decision tree (DT), and support vector machine (SVM). The LR model attained the highest testing accuracy at 99%, indicating stronger generalizability in discriminating between the LCS and HCS classes. While the DT model demonstrated excellent training accuracy, nonetheless, its test accuracy was noticeably lower at 93%, suggesting possible overfitting. The SVM exhibited more consistent behaviour, with 91% classification accuracy on both training and testing

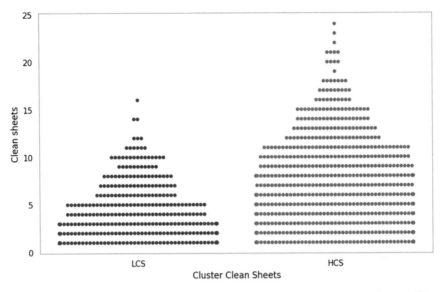

Fig. 3.1 Grouping of GKs clean sheets via k-means clustering. HCS = high clean sheets; LCS = low clean sheets

data. Further analysis of the confusion matrix for LR in Fig. 3.5 reveals the model's precise recognition capabilities for each class, given the input features. The LR matrix shows high precision and low misclassification rates overall. In summary, the empirical results demonstrate LR's superior predictive performance on unseen data for the present study. The visualization of model confusion matrices provides additional insight into the error characteristics of the LR model developed.

It is demonstrated from the findings of the current investigation that keeping clean sheets of GKs is associated with higher anthropometric indexes; specifically, the high clean sheets GKs are found to be heavier, taller, and older than the low clean sheets GKs. The fact that high clean sheets GKs are heavier and taller can be linked to their physical presence and shot-stopping abilities. Height gives GKs an advantage in reaching high shots and crosses, while weight can enhance their stability and ability to command their penalty area [17]. This combination makes it challenging for opponents to score, resulting in more clean sheets.

The observation that goalkeepers with a high percentage of clean sheets tend to be older suggests that the factors of experience and decision-making are critical to achieving success in goalkeeping. The advanced age of goalkeepers often equates to a greater wealth of experience in handling different match scenarios, thereby facilitating improved game reading abilities and the timely execution of decisions such as determining when to advance from the goal line, narrowing the angle, or maintaining a standing position [18]. On the other hand, the feet used by the GKs have no relationship with the ability of the GKs to keep clean sheets. This suggests that the type of dominant foot used by the GKs could not determine their clean sheet performances. The LR model applied in the current study has exhibited an excellent

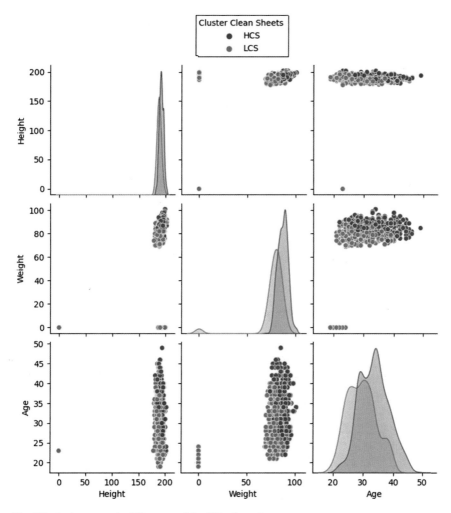

Fig. 3.2 Anthropometric differences of the GKs clean sheets group

classification ability in predicting the two groups of GKs. The efficacies of the LR model in solving a binary classification problem have been reported in the preceding investigations [13].

3.5 Summary

The present investigation has explored the association between the anthropometric index and clean sheets of GKs in European football tournaments. It was determined from the study findings that GKs with bigger body sizes and older are more likely to

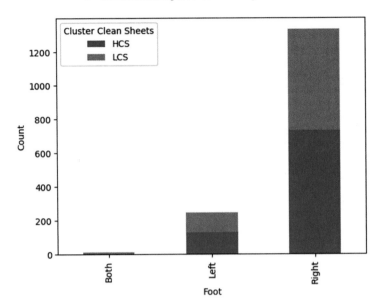

Fig. 3.3 Dominant foot and goalkeepers' clean sheets record performance

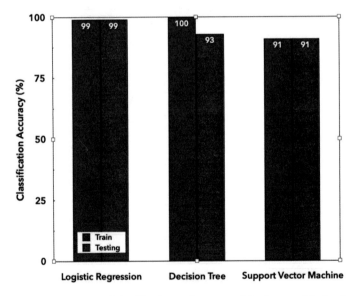

Fig. 3.4 Comparative analysis of classification performance of the three models developed

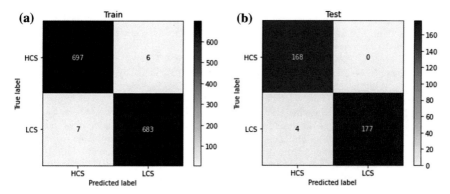

Fig. 3.5 Confusion matrix of the LR model: train **a** and test **b**

keep clean sheets in comparison with those with lower anthropometrics and younger. In addition, the findings revealed that the feet used by the GKs have no association with keeping clean sheets. Hence, higher anthropometric indexes and clean sheets in goalkeepers highlight the multidimensional nature of goalkeeping performance. It suggests that a combination of physical attributes as well as experience can contribute to a goalkeeper's ability to keep clean sheets. Understanding these factors can help teams make informed decisions in player recruitment and development which could go a long way in improving overall defensive performances and achieving greater success in football competitions. This should be considered while nurturing young goalkeepers to better equip them to cope with highly challenging goalkeeping tasks.

References

1. J. West, A review of the key demands for a football goalkeeper. Int. J. Sports Sci. Coach. **13**, 1215–1222 (2018)
2. M. Laurence, Want to compare keepers fairly? Take a deep dive into their clean sheets | Premier League | The Guardian. https://www.theguardian.com/football/who-scored-blog/2021/mar/10/want-to-compare-keepers-fairly-take-a-deep-dive-into-their-clean-sheets. Last accessed 13 Aug 2023
3. S.M. Gil, J. Zabala-Lili, I. Bidaurrazaga-Letona, B. Aduna, J.A. Lekue, J. Santos-Concejero, C. Granados, Talent identification and selection process of outfield players and goalkeepers in a professional soccer club. J. Sports Sci. **32**, 1931–1939 (2014)
4. V. Eswaramoorthi, M.R. Abdullah, R.M. Musa, A.B.H.M. Maliki, N.A. Kosni, N.B. Raj, N. Alias, H. Azahari, S.M. Mat-Rashid, H. Juahir, A multivariate analysis of cardiopulmonary parameters in archery performance. Hum. Mov. **19**, 35–41 (2018). https://doi.org/10.5114/hm.2018.77322
5. M.R. Abdullah, R.M. Musa, A. Maliki, P.K. Suppiah, N.A. Kosni, Relationship of physical characteristics, mastery and readiness to perform with position of elite soccer players. Int. J. Adv. Eng. Appl. Sci. **1**, 8–11 (2016)
6. M.R. Abdullah, A. Maliki, R.M. Musa, N.A. Kosni, H. Juahir, Intelligent prediction of soccer technical skill on youth soccer player's relative performance using multivariate analysis and artificial neural network techniques. Int. J. Adv. Sci. Eng. Inf. Technol. **6**, 668–674 (2016)

7. J. Torres-Unda, I. Zarrazquin, L. Gravina, J. Zubero, J. Seco, S.M. Gil, J. Gil, J. Irazusta, Basketball performance is related to maturity and relative age in elite adolescent players. J. Strength Cond. Res. **30**, 1325–1332 (2016)
8. R.M. Musa, A.P.P. Abdul Majeed, M.R. Abdullah, A.F.A.B. Nasir, M.H.A. Hassan, M.A.M. Razman, Technical and tactical performance indicators discriminating winning and losing team in elite Asian beach soccer tournament. PLoS One **14** (2019). https://doi.org/10.1371/journal.pone.0219138
9. M.R. Abdullah, V. Eswaramoorthi, R.M. Musa, A.B.H.M. Maliki, N.A. Kosni, M. Haque, The effectiveness of aerobic exercises at difference intensities of managing blood pressure in essential hypertensive information technology officers. J. Young Pharm. **8**, 483–486 (2016). https://doi.org/10.5530/jyp.2016.4.27
10. M.A. Gipit, M.R.A. Charles, R.M. Musa, N.A. Kosni, A.B.H.M. Maliki, The effectiveness of traditional games intervention programme in the improvement of form one school-age children's motor skills related performance components. Movement, Heal. Exerc. **6**, 157–169 (2017)
11. R. Muazu Musa, Z. Taha, A.P.P. Abdul Majeed, M.R. Abdullah, Psychological variables in ascertaining potential archers, in *SpringerBriefs in Applied Sciences and Technology* (2019), pp. 21–27. https://doi.org/10.1007/978-981-13-2592-2_3
12. R.M. Musa, A.P.P.A. Majeed, N.A. Kosni, M.R. Abdullah, *Machine Learning in Team Sports: Performance Analysis and Talent Identification in Beach Soccer & Sepak-takraw* (Springer Nature, 2020)
13. R. Muazu Musa, A.P.P. Abdul Majeed, M.R. Abdullah, G. Kuan, M.A. Mohd Razman, Current trend of analysis in high-performance sport and the recent updates in data mining and machine learning application in sports, in *Data Mining and Machine Learning in High-Performance Sport* (Springer, 2022), pp. 1–11
14. Z. Taha, M.A.M. Razman, F.A. Adnan, A.S. Abdul Ghani, A.P.P. Abdul Majeed, R.M. Musa, M.F. Sallehudin, Y. Mukai, The identification of hunger behaviour of lates calcarifer through the integration of image processing technique and support vector machine, in: *IOP Conference Series: Materials Science and Engineering* (2018). https://doi.org/10.1088/1757-899X/319/1/012028
15. M.A. Abdullah, M.A.R. Ibrahim, M.N.A. Bin Shapiee, M.A. Mohd Razman, R.M. Musa, A.P.P. Abdul Majeed, The classification of skateboarding trick manoeuvres through the integration of IMU and machine learning, in Lecture Notes in Mechanical Engineering (2020), pp. 67–74. https://doi.org/10.1007/978-981-13-9539-0_7
16. Z. Taha, R.M. Musa, A.P.P. Abdul Majeed, M.R. Abdullah, M.A. Zakaria, M.M. Alim, J.A.M. Jizat, M.F. Ibrahim, The identification of high potential archers based on relative psychological coping skills variables: a support vector machine approach, in *IOP Conference Series: Materials Science and Engineering* (2018). https://doi.org/10.1088/1757-899X/319/1/012027
17. H. Sarmento, M.T. Anguera, A. Pereira, D. Araújo, Talent identification and development in male football: a systematic review. Sport. Med. **48**, 907–931 (2018)
18. M. Mørch, Quantifying Footballing: Using Multiple Regression Analysis and ELO Ratings to Identify the Most Important KPI's for Goalkeepers (2020)

Chapter 4
A Machine Learning Analysis of Technical Skills and Tactical Awareness as Performance Predictors for Goalkeepers in European Football League

Abstract This chapter highlights the tactical and technical performance indicators that could distinguish between highly skilled and low-skilled goalkeepers (GKs). It has been demonstrated from the study findings that a set of indicators encompassing accurate and key passes, fouls, accurate foot passes from open play, accurate, hand passes, accurate passes from set pieces, short passes accurate, medium passes accurate, accurate long passes, and goals conceded could be essential in identifying the performance level of the GKs. It has also been demonstrated from the current finding that the application of the machine learning model is non-trivial in predicting the performance levels of the GKs. The decision tree model is found to yield an excellent prediction of the level of GKs levels with respect to the investigated indicators. It is then inferred that GKs' training programmes should include aspects related to said indicators.

Keywords Goalkeepers' performance · Machine learning models · European football leagues · Tactical awareness · Highly skilled goalkeepers

4.1 Overview

Goalkeepers (GKs) are an essential part of any football team, and their performance can often determine the outcome of a match. The role of a GK is to prevent the opposing team from scoring goals by using a combination of technical skills and tactical awareness [1]. Technical skills refer to the physical abilities required to make saves, such as reflexes, agility, and hand–eye coordination [2, 3]. Tactical awareness involves the ability to read the game, anticipate the movements of the opposing team, and position oneself accordingly.

Goalkeeping is a complex skill that requires a combination of technical, tactical, and physical abilities. In recent years, there has been a growing interest in the analysis of GK performance, with researchers seeking to identify the factors that contribute to success at the highest level. One of the most important factors in GK performance is technical skill [4]. GKs need to be able to catch, throw, and distribute the ball

R. M. Musa et al., *Data Mining and Machine Learning in Sports*,
SpringerBriefs in Applied Sciences and Technology,
https://doi.org/10.1007/978-981-99-7762-8_4

accurately, as well as make saves from a variety of shots. They also need to be able to command their area and communicate effectively with their defenders. Another important factor in GK's performance is tactical awareness. GKs need to be able to read the game and anticipate where shots are likely to come from. They also need to be able to make split-second decisions about whether to come off their line or stay on their line [5].

The European football league is one of the most competitive football leagues in the world, and the performance of GKs in this league is closely scrutinized. The league features some of the best GKs in the world, and their performances are often analysed to identify the factors that contribute to their success. This research paper aims to analyse the technical skills and tactical awareness of GKs in the European football league and their impact on performance.

4.2 Datasets and Treatment

In this study, we analysed a dataset of 1601 GKs from the top five European leagues (English Premier League, Spanish Laliga, Italian Serie A, French Ligue1, and German Bundesliga) across five consecutive seasons. The dataset included information on various technical and tactical indicators that consist of accurate passes, key passes, fouls, accurate foot passes from open play, accurate hand passes, accurate passes from set pieces, short passes accurate, medium passes accurate, accurate long passes, and goals conceded. Before the full analysis of this study began, the data underwent pre-processing and thorough checking for any missing information. Any rows that contained missing information were removed from the dataset [6, 7].

4.3 Clustering

The k-means clustering technique is a type of cluster analysis methodology that partitions a dataset into k predetermined and non-overlapping subgroups referred to as clusters, with each data point being assigned to only one cluster [8–10]. The approach endeavours to maximize the connectivity among intercluster data points, while simultaneously striving to maintain dissimilarity among intracluster data points to the greatest extent possible. The k-means cluster analysis was employed to partition GKs into two distinct groups based on the ten aforementioned indicators. Utilizing Euclidean distance as a metric, the two identified clusters were designated as high-skilled goalkeepers (HSG) and low-skilled goalkeepers (LSG).

4.4 Machine Learning-Based Classification Model

In this case study, an imbalance was observed in the distribution of classes, with 978 samples belonging to the LSG, while only 623 were from the HSG group. To mitigate the effects of this imbalance, the adaptive synthetic sampling approach (ADASYN) was employed to oversample the minority HSG class to 969 samples. It is worth noting that compared to random sampling, ADASYN reduces bias and improves learning by adaptively generating more useful synthetic data for the harder minority classes. The resultant balanced dataset was partitioned into an 80:20 ratio for training and testing sets, respectively [11–14]. Three supervised classification algorithms were implemented, namely logistic regression (LR), decision tree (DT), and support vector machine (SVM), to evaluate their efficacy in discriminating between the two classes. Overall classification accuracy and confusion matrix analysis were the evaluation metrics used. The programming environment consisted of the Spyder integrated development environment (IDE) version 5.4.3 with Python 3.9. Default hyperparameter configurations were utilized for the models without tuning. A comparative assessment was performed on the predictive capabilities of unseen test data. Both classifier precision and differential performance across classes were examined through generated confusion matrices.

4.5 Results and Discussion

Figure 4.1 depicts the differences between the GK classes identified through the k-means analysis with respect to the technical and tactical indicators examined. It could be seen from the figure that the HSG group recorded higher performances across all the indicators examined as compared to the LSG group.

The comparative classification performance of the three models—logistic regression (LR), decision tree (DT), and support vector machine (SVM)—is visualized in Fig. 4.2. The DT model attained the highest testing accuracy at 98%, indicating its stronger generalizability in discriminating between the low- and high-skilled GKs groups. The LR model achieved training and testing accuracies of 98% and 97%, respectively. Meanwhile, the SVM model attained an equal training and testing accuracy of 97%. Further analysis of the confusion matrix for the DT model in Fig. 4.3 reveals its precise recognition capabilities for each class given the input features. The DT matrix demonstrates overall high precision and low misclassification error rates. In summary, the empirical results indicate the DT model's superior predictive performance on unseen data in the present study. Visualization of the confusion matrices provides additional insights into the error characteristics of the developed DT model.

It is demonstrated from the findings of the current investigation that highly skilled GKs are distinguishable from the low skilled with respect to all the technical and tactical indicators examined. Several studies have been conducted to investigate the

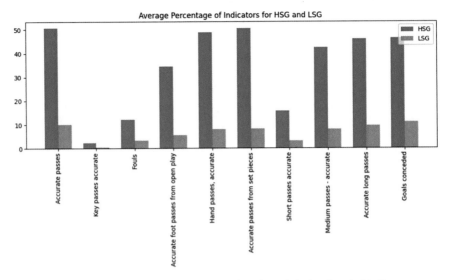

Fig. 4.1 Grouping of the goalkeepers' group based on the technical and tactical indicators

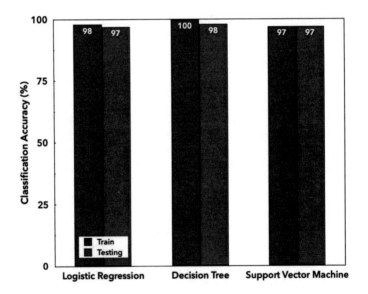

Fig. 4.2 Comparative performance of the three models examined

factors that contribute to the success of GKs in football. For example, a study by Sarmento et al. [15] found that technical skills such as diving, catching, and punching were essential for GKs to make successful saves. Another study by Lago-Peñas et al. [16] found that tactical awareness was a crucial factor in the performance of goalkeepers, as it allowed them to anticipate the movements of the opposing team

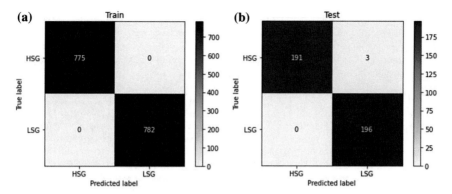

Fig. 4.3 Confusion matrix on the testing dataset

and position themselves accordingly. On the other hand, it is also observed that the highly skilled GKs are prone to conceive more goals compared to the low-skilled GKs. This might, however, be attributed to more games played by skilled GKs. The machine learning models applied in the current study have exhibited an excellent classification ability in predicting the two groups of GKs. It is determined from the study that the DT model outperformed the two models analysed. The efficacies of the DT model in solving a binary classification problem have been reported in the preceding investigations [17].

4.6 Summary

The present investigation has established tactical and technical indicators that could differentiate the performance levels of the GKs. These sets of actions included accurate passes, key passes accurate, fouls, accurate foot passes from open play, hand passes accurate, accurate passes from set pieces, short passes accurate, medium passes accurate, accurate long passes and goals conceded. It has also been demonstrated from the current finding that the DT model is able to yield an excellent prediction of the level of GKs with respect to investigated parameters. As a result, it is advised that the identification of high-performance GKs should consider the said tactical and technical indicators.

References

1. A. Welsh, *The Soccer Goalkeeping Handbook*, 3rd edn. (A&C Black, 2014)
2. M.R. Abdullah, R.M. Musa, A. Maliki, P.K. Suppiah, N.A. Kosni, Relationship of physical characteristics, mastery and readiness to perform with position of elite soccer players. Int. J. Adv. Eng. Appl. Sci. **1**, 8–11 (2016)

3. M.R. Abdullah, A. Maliki, R.M. Musa, N.A. Kosni, H. Juahir, Intelligent prediction of soccer technical skill on youth soccer player's relative performance using multivariate analysis and artificial neural network techniques. Int. J. Adv. Sci. Eng. Inf. Technol. **6**, 668–674 (2016)
4. C. Lago-Peñas, A. Dellal, Ball possession strategies in elite soccer according to the evolution of the match-score: the influence of situational variables. J. Hum. Kinet. **25**, 93–100 (2010)
5. M. Seaton, J. Campos, Distribution competence of a football clubs goalkeepers. Int. J. Perform. Anal. Sport **11**, 314–324 (2011)
6. M.R. Abdullah, V. Eswaramoorthi, R.M. Musa, A.B.H.M. Maliki, N.A. Kosni, M. Haque, The effectiveness of aerobic exercises at difference intensities of managing blood pressure in essential hypertensive information technology officers. J. Young Pharm. **8**, 483–486 (2016). https://doi.org/10.5530/jyp.2016.4.27
7. M.A. Gipit, M.R.A. Charles, R.M. Musa, N.A. Kosni, A.B.H.M. Maliki, The effectiveness of traditional games intervention programme in the improvement of form one school-age children's motor skills related performance components. Movement, Heal. Exerc. **6**, 157–169 (2017)
8. H. Azahari, H. Juahir, M.R. Abdullah, R.M. Musa, V. Eswaramoorthi, N. Alias, S.M. Mat-Rashid, N.A. Kosni, A.B.H.M. Maliki, N.B. Raj, A multivariate analysis of cardiopulmonary parameters in archery performance. Hum. Mov. **19**, 35–41 (2019). https://doi.org/10.5114/hm.2018.77322
9. M.R. Razali, N. Alias, A. Maliki, R.M. Musa, L.A. Kosni, H. Juahir, Unsupervised pattern recognition of physical fitness related performance parameters among Terengganu youth female field hockey players. Int. J. Adv. Sci. Eng. Inf. Technol. **7**, 100–105 (2017)
10. Z. Taha, M. Haque, R.M. Musa, M.R. Abdullah, A. Maliki, N. Alias, N.A. Kosni, Intelligent prediction of suitable physical characteristics toward archery performance using multivariate techniques. J Glob Pharma Technol. **9**, 44–52 (2009)
11. Z. Taha, M.A.M. Razman, F.A. Adnan, A.S. Abdul Ghani, A.P.P. Abdul Majeed, R.M. Musa, M.F. Sallehudin, Y. Mukai, The identification of hunger behaviour of lates calcarifer through the integration of image processing technique and support vector machine, in *IOP Conference Series: Materials Science and Engineering* (2018). https://doi.org/10.1088/1757-899X/319/1/012028
12. R.M. Musa, A.P.P. Abdul Majeed, M.R. Abdullah, A.F.A.B. Nasir, M.H.A. Hassan, M.A.M. Razman, Technical and tactical performance indicators discriminating winning and losing team in elite Asian beach soccer tournament. PLoS One. **14** (2019). https://doi.org/10.1371/journal.pone.0219138
13. M.A. Abdullah, M.A.R. Ibrahim, M.N.A. Bin Shapiee, M.A. Mohd Razman, R.M. Musa, A.P.P. Abdul Majeed, The classification of skateboarding trick manoeuvres through the integration of IMU and machine learning, in Lecture Notes in Mechanical Engineering (2020), pp. 67–74. https://doi.org/10.1007/978-981-13-9539-0_7
14. Z. Taha, R.M. Musa, A.P.P. Abdul Majeed, M.R. Abdullah, M.A. Zakaria, M.M. Alim, J.A.M. Jizat, M.F. Ibrahim, The identification of high potential archers based on relative psychological coping skills variables: a support vector machine approach, in *IOP Conference Series: Materials Science and Engineering* (2018). https://doi.org/10.1088/1757-899X/319/1/012027
15. H. Sarmento, R. Marcelino, M.T. Anguera, J. Campaniço, N. Matos, J.C. Leitão, Match analysis in football: a systematic review. J. Sports Sci. **32**, 1831–1843 (2014)
16. C. Lago-Peñas, The role of situational variables in analysing physical performance in soccer. J. Hum. Kinet. **35**, 89 (2012)
17. R. Muazu Musa, A.P.P. Abdul Majeed, M.R. Abdullah, G. Kuan, M.A. Mohd Razman, Current trend of analysis in high-performance sport and the recent updates in data mining and machine learning application in sports, in *Data Mining and Machine Learning in High-Performance Sport* (Springer, 2022), pp. 1–11

Chapter 5
An Unsupervised Analysis of Identifying Key Technical and Tactical Skills Awareness Among Top European Goalkeepers

Abstract This chapter highlights the crucial tactical and technical-related indicators that are vital towards the successful performance of goalkeepers (GKs). These indicators, which consist of good interception, accurate passes, accurate foot passes from open play, hand passes accurate, accurate passes from set pieces, medium passes accurate, and accurate long passes, are deemed significant in determining the success of goalkeepers in the European football championship. Furthermore, the utilization of principal component analysis has proven to be effective in extracting the most important indicators that can explain successful GKs' performance. Therefore, it is recommended that these highlighted indicators be incorporated into the training routine of GKs. This is particularly important when nurturing young GKs to better prepare them to handle high-pressure goalkeeping tasks.

Keywords Goalkeepers · Training strategy · Tactical awareness · Principal component analysis · European football league

5.1 Overview

Goalkeepers (GKs) are considered to be among the most crucial players in the game of football, owing to their ability to prevent goals and initiate attacks. However, achieving excellence in this position requires more than just physical prowess. GKs must also develop a keen sense of technical and tactical skills [1]. In football, technical skills refer to the physical abilities required to perform the sport, including passing, shooting, dribbling, and saving. Tactical skills, on the other hand, encompass the mental abilities necessary to make effective decisions and execute strategies, such as positioning, communication, anticipation, and leadership [2–4]. Given their unique role in the game, GKs must possess both types of skills to succeed.

© The Author(s), under exclusive license to Springer Nature Singapore Pte Ltd. 2024   35
R. M. Musa et al., *Data Mining and Machine Learning in Sports*,
SpringerBriefs in Applied Sciences and Technology,
https://doi.org/10.1007/978-981-99-7762-8_5

The position of the GKs in football is complex and challenging, necessitating a combination of physical, technical, and tactical abilities. The physical abilities include power, swiftness, dexterity, and synchronization. The technical abilities encompass handling, footwork, distribution, and reflexes. The tactical abilities involve perception, judgement, arrangement, and communication [5].

The European football league is widely regarded as one of the most fiercely competitive football leagues worldwide, and as such, the performance of GKs in this league is subject to intense scrutiny. This league features some of the finest GKs globally, and their performances are meticulously analysed to identify the key factors that contribute to their success. The objective of this investigation is to examine the technical abilities and tactical awareness of goalkeepers in the European football league and their influence on their performance using an unsupervised technique.

5.2   Data Source and Treatment

In this study, we analysed a dataset of 1601 goalkeepers from the top five European leagues (English Premier League, Spanish Laliga, Italian Serie A, French Ligue1, and German Bundesliga) across five consecutive seasons. The dataset included information on various technical and tactical indicators that consist of good interception of GKs, accurate passes, key passes accurate, fouls, accurate foot passes from open play, hand passes accurate, accurate passes from set pieces, short passes accurate, medium passes accurate, and accurate long passes. Before the full analysis in this study began, the data underwent pre-processing and thorough checking for any missing information. Any rows that contained missing information were removed from the dataset [6, 7].

5.3   Feature Selection

In the present study, a principal component analysis (PCA) was employed in order to ascertain the essential tactical and technical indicators through the extraction of the most pertinent actions [8–10]. It is important to note that the process of extracting data from the PCA involves the elimination of the data that constitutes the least significant component, followed by the retention of the most useful information in the data [11, 12]. The statistical analysis was performed with the aid of the XLSTAT2014 add-in software for Windows.

5.4 Results and Discussion

Figure 5.1 depicts the scree plot of the eigenvalues obtained from the PCA. It is evident from the figure that a single component could account for the essential indicators. The component constitutes tactical and technical actions that are deemed most important owing to their relatively higher eigenvalues (greater than 1). These components were identified and retained for use as input parameters in subsequent analyses, specifically varimax rotation.

Table 5.1 presents the results of the PCA with varimax rotation. It is evident from the table that within the identified component, some relevant and related tactical and technical indicators are identified. These indicators have been identified based on their fulfilment of the preset factor loading threshold, which is equal to or greater than 0.80. Moreover, it can be observed that 7 out of the 10 initially examined technical and tactical actions were identified as most essential for the GKs in the top European leagues' tournaments.

It is demonstrated from the findings of the current investigation that technical and tactical skills parameters, namely good interception, accurate passes, accurate foot passes from open play, hand passes accurate, accurate passes from set pieces, medium passes accurate, and accurate long passes, are most essential for GKs performance in the European football leagues tournament. These skills enable the GKs to play a fundamental tactical role at the base of the team, initiate attacks with precise distribution, and handle different types of shots with confidence [13]. According to a recent study, elite goalkeepers in professional soccer perform an average of 36.5

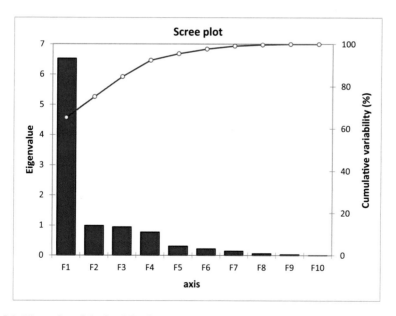

Fig. 5.1 Eigenvalue of the first PCA for components extractions

Table 5.1 Principal component analysis results after rotation

Variables	Factor loadings	Contribution (%)
Good interception of goalkeeper	**0.895**	12.279
Accurate passes	**0.982**	14.793
Key passes accurate	0.284	1.238
Fouls	0.387	2.301
Accurate foot passes from open play	**0.928**	13.203
Hand passes accurate	**0.918**	12.921
Accurate passes from set pieces	**0.894**	12.265
Short passes accurate	0.727	8.109
Medium passes accurate	**0.912**	12.741
Accurate long passes	**0.814**	10.150
Eigenvalue	*6.523*	
Cumulative (%)	65.229	*100.000*

Note The number in bold indicates an important variable

technical–tactical actions per match, with passing being the most frequent action (48.8%), followed by catching (18.5%), and diving (10.3%) [14]. Therefore, it is crucial for GKs to master these skills and apply them effectively in different game situations.

5.5   Summary

In the present investigation, we identified the necessary tactical and technical-related indicators that are essential towards GK performances. These sets of indicators that included good interception, accurate passes, accurate foot passes from open play, hand passes accurate, accurate passes from set pieces, medium passes accurate, and accurate long passes are found to be non-trivial in determining the successful performance of GKs in the European football championship. It has also been demonstrated from the current finding that the application of principal component analysis serves useful in extracting the most important indicators that could explain successful GKs performance. As a result, it is advised that GK training should incorporate the highlighted indicators as part of the GK training routine. This should be considered while nurturing young GKs to better equip them to cope with highly challenging goalkeeping tasks.

References

1. A. White, S.P. Hills, C.B. Cooke, T. Batten, L.P. Kilduff, C.J. Cook, C. Roberts, M. Russell, Match-play and performance test responses of soccer goalkeepers: a review of current literature. Sport. Med. **48**, 2497–2516 (2018)
2. M. Seaton, J. Campos, Distribution competence of a football clubs goalkeepers. Int. J. Perform. Anal. Sport **11**, 314–324 (2011)
3. R.M. Musa, A.P.P. Abdul Majeed, M.R. Abdullah, A.F.A.B. Nasir, M.H.A. Hassan, M.A.M. Razman, Technical and tactical performance indicators discriminating winning and losing team in elite Asian beach soccer tournament. PLoS One. **14** (2019). https://doi.org/10.1371/journal. pone.0219138
4. M.R. Abdullah, A. Maliki, R.M. Musa, N.A. Kosni, H. Juahir, Intelligent prediction of soccer technical skill on youth soccer player's relative performance using multivariate analysis and artificial neural network techniques. Int. J. Adv. Sci. Eng. Inf. Technol. **6**, 668–674 (2016)
5. F. Santos, J. Santos, M. Espada, C. Ferreira, P. Sousa, V. Pinheiro, T-pattern analysis of offensive and defensive actions of youth football goalkeepers. Front. Psychol. 5371 (2022)
6. M.R. Abdullah, V. Eswaramoorthi, R.M. Musa, A.B.H.M. Maliki, N.A. Kosni, M. Haque, The effectiveness of aerobic exercises at difference intensities of managing blood pressure in essential hypertensive information technology officers. J. Young Pharm. **8**, 483–486 (2016). https://doi.org/10.5530/jyp.2016.4.27
7. M.A. Gipit, M.R.A. Charles, R.M. Musa, N.A. Kosni, A.B.H.M. Maliki, The effectiveness of traditional games intervention programme in the improvement of form one school-age children's motor skills related performance components. Movement, Heal. Exerc. **6**, 157–169 (2017)
8. Z. Taha, M. Haque, R.M. Musa, M.R. Abdullah, A. Maliki, N. Alias, N.A. Kosni, Intelligent prediction of suitable physical characteristics toward archery performance using multivariate techniques. J Glob Pharma Technol. **9**, 44–52 (2009)
9. M.R. Abdullah, A.B.H.M. Maliki, R.M. Musa, N.A. Kosni, H. Juahir, S.B. Mohamed, Identification and comparative analysis of essential performance indicators in two levels of soccer expertise. Int. J. Adv. Sci. Eng. Inf. Technol. **7**, 305–314 (2017). https://doi.org/10.18517/ija seit.7.1.1150
10. V. Eswaramoorthi, M.R. Abdullah, R.M. Musa, A.B.H.M. Maliki, N.A. Kosni, N.B. Raj, N. Alias, H. Azahari, S.M. Mat-Rashid, H. Juahir, A multivariate analysis of cardiopulmonary parameters in archery performance. Hum. Mov. **19**, 35–41 (2018). https://doi.org/10.5114/hm. 2018.77322
11. M.R. Abdullah, R.M. Musa, N.A. Kosni, A. Maliki, M. Haque, Profiling and distinction of specific skills related performance and fitness level between senior and junior Malaysian youth soccer players. Int. J. Pharm. Res. **8**, 64–71 (2016)
12. M.R. Razali, N. Alias, A. Maliki, R.M. Musa, L.A. Kosni, H. Juahir, Unsupervised pattern recognition of physical fitness related performance parameters among Terengganu youth female field hockey players. Int. J. Adv. Sci. Eng. Inf. Technol. **7**, 100–105 (2017)
13. The Complete Goalkeeper Training Guide (Drills and Tips), https://www.soccercoachingpro. com/goalkeeper-training/. Last accessed 13 Aug 2023
14. M. Obetko, P. Peráček, M. Mikulič, M. Babic, Technical–tactical profile of an elite soccer goalkeeper. J. Phys. Educ. Sport. **22**, 38–46 (2022)

Chapter 6
Identification of Dominant Types of Shots Received and Saves Among Top European Goalkeepers: A Significant Attribute Evaluation Technique

Abstract This chapter highlights the dominant shots and saves directed at goal-keepers (GKs) in the European football championship. It has been demonstrated from the study findings that a set of actions encompassing stopped shots, saves without jumping, jumping saves, shots saved with unsuccessful bouncing, shots saved with successful bouncing as well as close-range shots on target are highly dominant for GKs performances in the European football tournament. It is then inferred that training regimes and tactical strategies should be considered to improve on the weaknesses and leverage the strengths of the GKs.

Keywords Goalkeepers' performances · Training strategy · Goalkeepers' techniques · Feature importance analysis · European football league

6.1 Overview

The role of the goalkeeper (GKs) in football has evolved over time, from being a passive defender to being an active participant in the game. The GKs not only have to stop shots from the opponents, but also have to distribute the ball effectively and communicate with the teammates [1]. However, the evaluation of goalkeeper performance is often based on simple metrics such as goals conceded or clean sheets, which do not reflect the complexity and diversity of GK actions.

It is worth highlighting that goalkeeping is a complex and demanding skill that requires physical, technical, tactical, and psychological abilities [2–4]. The performance of a GK can have a significant impact on the outcome of a football match, as well as on the confidence and morale of the team [5].

GKs play a pivotal role in the game of football, often being the last line of defence for their teams. Their ability to make crucial saves can turn the tide of a match and significantly impact the outcome of a game [6]. In the context of top-tier European football tournaments, where competition is fierce and margins for error are slim, understanding the dominant types of shots received by GKs and the corresponding saves they make becomes imperative for player development and team tactics. Hence,

© The Author(s), under exclusive license to Springer Nature Singapore Pte Ltd. 2024   41
R. M. Musa et al., *Data Mining and Machine Learning in Sports*,
SpringerBriefs in Applied Sciences and Technology,
https://doi.org/10.1007/978-981-99-7762-8_6

the current investigation aims to analyse and identify the most prevalent types of shots faced by top European GKs and to explore the techniques the GKs employ to save these shots.

6.2 Data Source and Treatment

In the present study, a dataset comprising 1601 goalkeepers from the top five European leagues, namely English Premier League, Spanish Laliga, Italian Serie A, French Ligue1, and German Bundesliga, across five consecutive seasons was analysed. This dataset comprised several shots and saves indicators directed to the GKs including stopped shots, saves without jumping, jumping saves, shots saved with unsuccessful bouncing, shots saved with successful bouncing, close-range shots on target, mid-range shots, close-range shots saved, close-range shots, mid-range shots on target, mid-range shots saved, long-range shots saved, long-range shots on target, and long-range shots. It is worth noting that before commencing with the full analysis, the data underwent a pre-processing phase and was thoroughly checked for any missing information. Any rows that contained missing information were subsequently removed from the dataset [7, 8].

6.3 A Significant Attribute Evaluation Technique

Significant Attribute Evaluation Technique (SAET) is a statistical technique that can be utilized to ascertain the most crucial factors in a given dataset. This technique employs a filter methodology, which implies that it does not necessitate the utilization of a learning algorithm. It is unique in comparison with other filter methods since it employs a hypothesis testing approach to identify significant factors, can be employed to recognize interaction effects, and is a nonparametric method. The technique goes beyond prioritizing attributes that have a significant impact on the overall performance of a model. SAET employs various statistical measures and algorithms to assess the relevance of each attribute in relation to the target variable, aiming to retain only the most informative and discriminative features. Consequently, SEAT is a potent and versatile method for identifying significant factors in a given dataset, particularly for identifying interaction effects and dealing with nonparametric data. In this investigation, we applied SEAT to identify the most essential shots received and saves performed by the GKs.

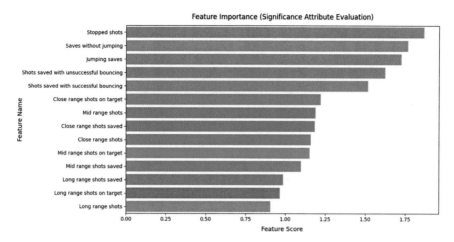

Fig. 6.1 Most influential saves and shots identified via feature importance analysis

6.4 Results and Discussion

Figure 6.1 depicts the analysis of the significance attribute evaluation. It could be seen from the figure that certain types of shots and saves of the GKs are displayed as most crucial. These indicators consist of stopped shots, saves without jumping, jumping saves, shots saved with unsuccessful bouncing, shots saved with successful bouncing as well as close-range shots on target.

It is demonstrated from the findings of the current investigation that actions that encapsulate stopped shots, save without jumping, jumping saves, shots saved with unsuccessful bouncing, shots saved with successful bouncing as well as close-range shots on target are highly dominant for GK performances in the European football tournament. In other words, GKs are most frequently faced with the aforesaid actions during European football competitions. Being aware of the most frequent types of actions that GKs face can be valuable for coaches, players, and analysts to understand goalkeeper performance trends and work on specific skills and strategies to improve their effectiveness in these situations [9–11].

6.5 Summary

The present investigation has established the dominant saves and shots directed at goalkeepers in the European football championship. From the study's findings, it has been demonstrated that a series of actions, including stopped shots, save without jumping, jumping saves, shots saved with unsuccessful bouncing, shots saved with successful bouncing, as well as close-range shots on target, is of utmost importance for the goalkeepers' performances in the European football tournament. As such, it is inferred that training regimes and tactical strategies ought to be duly considered to enhance the weaknesses and leverage the strengths of the goalkeepers.

References

1. H. Liu, M.A. Gómez, C. Lago-Peñas, Match performance profiles of goalkeepers of elite football teams. Int. J. Sports Sci. Coach. **10**, 669–682 (2015)
2. M.R. Abdullah, R.M. Musa, A. Maliki, P.K. Suppiah, N.A. Kosni, Relationship of physical characteristics, mastery and readiness to perform with position of elite soccer players. Int. J. Adv. Eng. Appl. Sci. **1**, 8–11 (2016)
3. M.R. Abdullah, A. Maliki, R.M. Musa, N.A. Kosni, H. Juahir, Intelligent prediction of soccer technical skill on youth soccer player's relative performance using multivariate analysis and artificial neural network techniques. Int. J. Adv. Sci. Eng. Inf. Technol. **6**, 668–674 (2016)
4. R.M. Musa, A.P.P. Abdul Majeed, M.R. Abdullah, A.F.A.B. Nasir, M.H.A. Hassan, M.A.M. Razman, Technical and tactical performance indicators discriminating winning and losing team in elite Asian beach soccer tournament. PLoS One. **14** (2019). https://doi.org/10.1371/journal.pone.0219138
5. P. Sainz de Baranda, L. Adán, A. García-Angulo, M. Gómez-López, B. Nikolic, E. Ortega-Toro, Differences in the offensive and defensive actions of the goalkeepers at women's FIFA World Cup 2011. Front. Psychol. **10**, 223 (2019)
6. A. Szwarc, J. Jaszczur-Nowicki, P. Aschenbrenner, M. Zasada, J. Padulo, P. Lipinska, Motion analysis of elite Polish soccer goalkeepers throughout a season. Biol. Sport **36**, 357–363 (2019)
7. M.R. Abdullah, V. Eswaramoorthi, R.M. Musa, A.B.H.M. Maliki, N.A. Kosni, M. Haque, The effectiveness of aerobic exercises at difference intensities of managing blood pressure in essential hypertensive information technology officers. J. Young Pharm. **8**, 483–486 (2016). https://doi.org/10.5530/jyp.2016.4.27
8. M.A. Gipit, M.R.A. Charles, R.M. Musa, N.A. Kosni, A.B.H.M. Maliki, The effectiveness of traditional games intervention programme in the improvement of form one school-age children's motor skills related performance components. Movement, Heal. Exerc. **6**, 157–169 (2017)
9. R. Muazu Musa, A.P.P. Abdul Majeed, M.R. Abdullah, G. Kuan, M.A. Mohd Razman, Current trend of analysis in high-performance sport and the recent updates in data mining and machine learning application in sports, in *Data Mining and Machine Learning in High-Performance Sport* (Springer, 2022), pp. 1–11
10. V. Eswaramoorthi, M.R. Abdullah, R.M. Musa, A.B.H.M. Maliki, N.A. Kosni, N.B. Raj, N. Alias, H. Azahari, S.M. Mat-Rashid, H. Juahir, A multivariate analysis of cardiopulmonary parameters in archery performance. Hum. Mov. **19**, 35–41 (2018). https://doi.org/10.5114/hm.2018.77322
11. M. Dicks, C. Button, K. Davids, Examination of gaze behaviors under in situ and video simulation task constraints reveals differences in information pickup for perception and action. Attention, Perception, Psychophys. **72**, 706–720 (2010)

Chapter 7
Evaluation of Goalkeepers' Goals Conceptions from Different Saves and Shots Indicators

Abstract This chapter highlights various shots and saves as well as their association with goalkeepers' (GKs) goal conceptions in European championship tournaments. It has been demonstrated from the study findings that a set of shots and saves encompassing mid-range shots saved, close-range saves, mid-range shots saved, and long-range shots are vital predictors of GK's goals conceptions. It has also been demonstrated from the current finding that the logistic regression model is able to yield an excellent prediction of the GKs grouping with respect to the investigated parameters. It is then inferred that GKs' training programmes should include aspects related to dealing with various shots and specific saves necessary during match play.

Keywords Logistic regression model · Feature importance analysis · Goalkeepers performance · Goals conception · European Football league

7.1 Overview

Football, being a sport with a high level of unpredictability, demands GKs to possess exceptional skills to prevent their opponents from putting the ball into the back of the net. To conduct a comprehensive evaluation of a GK's goals conceptions, it is essential to have a thorough understanding of the concept of saves. Saves can be described as instances where a GK effectively thwarts the opposing team's attempt to score a goal. While this definition may seem straightforward, it is important to acknowledge the complexities involved in each save. During the process of making a save, a GK must anticipate the actions and movements of the opposing team members, position themselves most optimally, and execute the appropriate technique to prevent a goal [1]. These factors present significant challenges, necessitating that GKs possess a combination of technical skills, decision-making abilities, and tactical awareness.

Another crucial factor to consider while evaluating the GK's goals conceptions is the examination of shots. This involves the attempts made by the opposing team to score a goal. Shots can differ in their intensity, trajectory, and distance from the goal, thereby presenting unique challenges to the GKs [2–4]. All shots are not created equally, and some require exceptional reflexes, acrobatic ability, or strategic positioning to prevent a goal [5]. The analysis of various types of shots and their conversion rates provides valuable insights into the effectiveness of GKs. Furthermore, the study of shots helps in identifying patterns and trends, facilitating the development of effective defensive strategies. By comprehending the interplay between saves and shots, one can unearth invaluable information about the GKs' goals conceptions in diverse game scenarios.

In this study, an examination was carried out on a dataset consisting of 1325 GKs from the top five European leagues—English Premier League, Spanish Laliga, Italian Serie A, French Ligue1, and German Bundesliga, throughout five consecutive seasons. This dataset included various indicators for shots and saves aimed at the GKs, such as stopped shots, saves without jumping, jumping saves, shots saved with unsuccessful bouncing, shots saved with successful bouncing, close-range shots on target, mid-range shots, close-range shots saved, close-range shots, mid-range shots on target, mid-range shots saved, long-range shots saved, long-range shots on target, and long-range shots. It should be noted that prior to beginning the complete analysis, the data underwent a pre-processing phase and was meticulously examined for any missing information. Any rows that contained missing information were subsequently eliminated from the dataset [6, 7].

7.2 Clustering

Hierarchical agglomerative cluster (HACA) analysis serves both as an exploratory tool and a non-exploratory technique. It establishes a cluster hierarchy for a single observation and groups related observations into distinct observations [8]. It is worth noting that in this algorithm, the learning process is driven by both data set merges and splits [9–11]. The HACA was used to assign the GK goals conceptions into groups based on the number of frequencies across the games played. In this analysis, the cosine distance was utilized, and the clustering validation technique was carried out using class centroids [41] for the formation of the two detected clusters, namely less goals conceived GKs (LGC) and more goals conceived (MGC).

7.3 Machine Learning-Based Classification Model

In this case study, a total of 1325 samples were obtained of which 763 belong to the LGC, while the remaining 562 belong to the MGC class. Although there is a slight imbalance, however, with the ratio of 1.4:1, it could be treated as is. The dataset was split into an 80:20 ratio for training and testing sets, respectively [12–14]. In this study, we aim to investigate the importance of these features. Therefore, only the logistic regression (LR) utilized to evaluate its efficacy in discriminating between the two classes by considering all features and features selected by the random forest (based on the Gini Importance) model. The classification accuracy (CA) and confusion matrix analysis were used as the evaluation metrics. The Spyder integrated development environment (IDE) version 5.4.3 was used as the programming environment that runs on Python 3.9. Default hyperparameter configurations of the LR model were utilized in this study.

7.4 Results and Discussion

Figure 7.1 illustrates the categories of goalkeepers, ascertained through HACA analysis, based on the number of goals conceded. The figure clearly indicates a distinct separation between the LGC and MGC. In other words, the average frequency of goals conceded by the GKs in the MGC is significantly greater than that of the LGC.

Figure 7.2 projects the variations of the GKs with regard to all the indicators examined. It could be observed from the figure that GKs who conceived more goals possessed a higher degree of shots and saves, while the less goals conceived GKs were confronted with fewer shots and recorded fewer saves. Higher activity observed from the MGC GKs could be attributed to more games played in comparison with the LGC GKs. These findings reflect that higher games played by GKs are associated with more goal conceptions.

Figure 7.3 illustrates the performance comparison of the logistic regression (LR) model using all features versus selected features. Considering all features resulted in training and testing classification accuracies (CA) of 89% and 88%, respectively. In contrast, employing the selected features identified via the random forest (RF) model in Fig. 7.4, namely mid-range shots saved (MRS), close-range saves (CRS), mid-range shots saved (MROS), and long-range shots (LRS), yielded markedly higher accuracies on both the training and testing datasets with CAs of 97% and 98%,

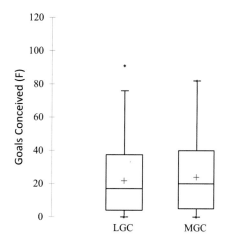

Fig. 7.1 Grouping of goalkeepers' goals conception via HACA clustering. LGC = less goals conceived; MGC = more goals conceived

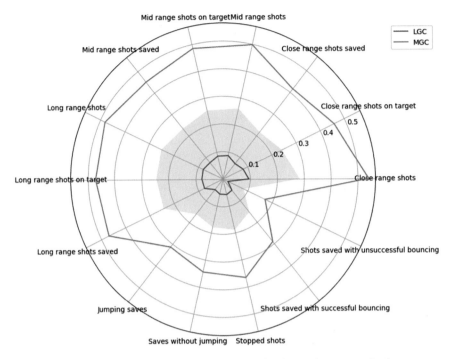

Fig. 7.2 Differences between the goalkeepers based on the shots and saves examined

respectively. This finding indicates the LR model attains superior predictive performance when restricted to significant features, while non-selected features may be
extraneous and detrimental to the model. Furthermore, Fig. 7.5 provides additional
insights into the model misclassification tendencies when employing all features
versus selected features. The comparative analysis demonstrates that feature selection serves to improve generalization and avoid overfitting for the LR classifier in
this study [15].

It is demonstrated from the findings of the current investigation that a set of shots
and saves encompassing mid-range shots saved, close-range saves, mid-range shots
saved as well as long-range shots are vital predictors for GK's goal conceptions.
In modern football, attackers are becoming more proficient at taking shots from
various positions. As a result, GKs must be prepared to face shots from different
distances throughout a game [16, 17]. A GK's ability to save mid-range shots, close-
range efforts, and long-range strikes helps them adapt to the dynamic nature of
the game. Moreover, GKs who can successfully save shots from different distances
demonstrate a well-rounded skill set. Mid-range shots represent a common occur-
rence during a match, while close-range saves are crucial for thwarting point-blank
attempts by opposition players. On the other hand, long-range shots require excellent
positioning, agility, and reflexes to prevent unexpected goals [18, 19]. The current

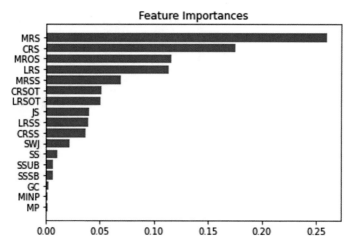

Fig. 7.3 Significant features identified by the RF (Gini Importance)

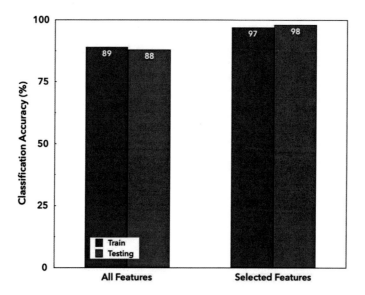

Fig. 7.4 Comparison between considering all features and selected features based on the LR model

investigation's findings strongly indicate that a goalkeeper's ability to handle mid-range shots, close-range saves, and long-range shots is a vital predictor for their effectiveness in preventing goals. It has been previously reported that diverse shot and save repertoire not only enhance a GK's overall performance but also contribute significantly to their team's defensive success [20–22].

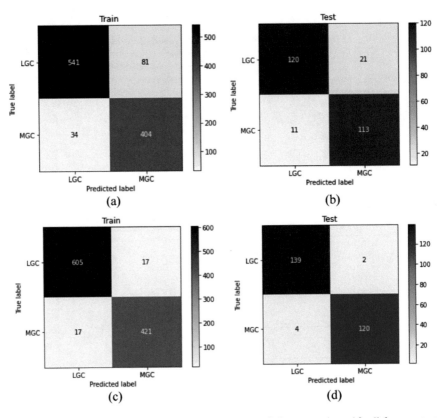

Fig. 7.5 Confusion matrix of the LR model based on **a** all features train and **b** all features test. **a** Selected features train and **b** selected features test

7.5 Summary

The present investigation has determined different shots and saves and their connection with goalkeepers, particularly in regard to goals conceived during European championship tournaments. The study findings have demonstrated that a specific set of shots and saves, including mid-range and long-range shots saved, as well as close-range saves, are essential predictors of GKs' goals conceived. Additionally, the current findings have shown that the logistic regression model can accurately predict the GKs' grouping based on the investigated parameters. Therefore, it can be inferred that GKs' training programmes should incorporate strategies for dealing with various shots and specific saves required during match play. Future research in this field should incorporate both quantitative and qualitative aspects which could consequently play a vital role in enhancing our understanding and assessment of GKs' goals conceptions, leading to improved training methodologies and performance analysis techniques in the world of football.

References

1. B. Noël, J. Van der Kamp, S. Klatt, The interplay of goalkeepers and penalty takers affects their chances of success. Front. Psychol. **12**, 645312 (2021)
2. M.R. Abdullah, R.M. Musa, A. Maliki, P.K. Suppiah, N.A. Kosni, Relationship of physical characteristics, mastery and readiness to perform with position of elite soccer players. Int. J. Adv. Eng. Appl. Sci. **1**, 8–11 (2016)
3. M.R. Abdullah, A. Maliki, R.M. Musa, N.A. Kosni, H. Juahir, Intelligent prediction of soccer technical skill on youth soccer player's relative performance using multivariate analysis and artificial neural network techniques. Int. J. Adv. Sci. Eng. Inf. Technol. **6**, 668–674 (2016)
4. R.M. Musa, A.P.P. Abdul Majeed, M.R. Abdullah, A.F.A.B. Nasir, M.H.A. Hassan, M.A.M. Razman, Technical and tactical performance indicators discriminating winning and losing team in elite Asian beach soccer tournament. PLoS One. **14** (2019). https://doi.org/10.1371/journal.pone.0219138
5. G. Gelade, Evaluating the ability of goalkeepers in English Premier League football. J. Quant. Anal. Sport. **10**, 279–286 (2014)
6. M.R. Abdullah, V. Eswaramoorthi, R.M. Musa, A.B.H.M. Maliki, N.A. Kosni, M. Haque, The effectiveness of aerobic exercises at difference intensities of managing blood pressure in essential hypertensive information technology officers. J. Young Pharm. **8**, 483–486 (2016). https://doi.org/10.5530/jyp.2016.4.27
7. M.A. Gipit, M.R.A. Charles, R.M. Musa, N.A. Kosni, A.B.H.M. Maliki, The effectiveness of traditional games intervention programme in the improvement of form one school-age children's motor skills related performance components. Movement, Heal. Exerc. **6**, 157–169 (2017)
8. O. Maimon, L. Rokach, Data Mining and Knowledge Discovery Handbook (2005). https://doi.org/10.1007/b107408
9. A.B.H.M. Maliki, M.R. Abdullah, H. Juahir, F. Abdullah, N.A.S. Abdullah, R.M. Musa, S.M. Mat-Rasid, A. Adnan, N.A. Kosni, W.S.A.W. Muhamad, N.A.M. Nasir, A multilateral modelling of Youth Soccer Performance Index (YSPI). IOP Conf. Ser. Mater. Sci. Eng. **342**, 012057 (2018). https://doi.org/10.1088/1757-899X/342/1/012057
10. M.R. Razali, N. Alias, A. Maliki, R.M. Musa, L.A. Kosni, H. Juahir, Unsupervised pattern recognition of physical fitness related performance parameters among Terengganu youth female field hockey players. Int. J. Adv. Sci. Eng. Inf. Technol. **7**, 100–105 (2017)
11. Z. Taha, R.M. Musa, M.R. Abdullah, A. Maliki, N.A. Kosni, S.M. Mat-Rasid, A. Adnan, H. Juahir, Supervised pattern recognition of archers' relative psychological coping skills as a component for a better archery performance. J. Fundam. Appl. Sci. **10**, 467–484 (2018)
12. Z. Taha, R.M. Musa, A.P.P. Abdul Majeed, M.R. Abdullah, M.A. Zakaria, M.M. Alim, J.A.M. Jizat, M.F. Ibrahim, The identification of high potential archers based on relative psychological coping skills variables: a support vector machine approach, in *IOP Conference Series: Materials Science and Engineering* (2018). https://doi.org/10.1088/1757-899X/319/1/012027
13. M.A. Abdullah, M.A.R. Ibrahim, M.N.A. Bin Shapiee, M.A. Mohd Razman, R.M. Musa, A.P.P. Abdul Majeed, The classification of skateboarding trick manoeuvres through the integration of IMU and machine learning, in Lecture Notes in Mechanical Engineering (2020), pp. 67–74. https://doi.org/10.1007/978-981-13-9539-0_7
14. Z. Taha, M.A.M. Razman, F.A. Adnan, A.S. Abdul Ghani, A.P.P. Abdul Majeed, R.M. Musa, M.F. Sallehudin, Y. Mukai, The identification of hunger behaviour of lates calcarifer through the integration of image processing technique and support vector machine, in *IOP Conference Series: Materials Science and Engineering* (2018). https://doi.org/10.1088/1757-899X/319/1/012028
15. R. Muazu Musa, A.P.P. Abdul Majeed, M.R. Abdullah, G. Kuan, M.A. Mohd Razman, Current trend of analysis in high-performance sport and the recent updates in data mining and machine learning application in sports, in *Data Mining and Machine Learning in High-Performance Sport* (Springer, 2022), pp. 1–11

16. J. West, A review of the key demands for a football goalkeeper. Int. J. Sports Sci. Coach. **13**, 1215–1222 (2018)
17. M. Mørch, Quantifying Footballing: Using Multiple Regression Analysis and ELO Ratings to Identify the Most Important KPI's for Goalkeepers (2020)
18. A. Welsh, *The Soccer Goalkeeping Handbook*, 3rd edn. (A&C Black, 2014)
19. M. Seaton, J. Campos, Distribution competence of a football clubs goalkeepers. Int. J. Perform. Anal. Sport **11**, 314–324 (2011)
20. H. Liu, M.A. Gómez, C. Lago-Peñas, Match performance profiles of goalkeepers of elite football teams. Int. J. Sports Sci. Coach. **10**, 669–682 (2015)
21. A. Szwarc, J. Jaszczur-Nowicki, P. Aschenbrenner, M. Zasada, J. Padulo, P. Lipinska, Motion analysis of elite Polish soccer goalkeepers throughout a season. Biol. Sport **36**, 357–363 (2019)
22. F. Santos, J. Santos, M. Espada, C. Ferreira, P. Sousa, V. Pinheiro, T-pattern analysis of offensive and defensive actions of youth football goalkeepers. Front. Psychol. 5371 (2022)

Chapter 8
Summary, Conclusion, Current Status, and Future Direction for Goalkeeper's Performance in European Football Leagues

Abstract In this chapter, we summarized the major findings of the study. The status of the goalkeeper's performance in the European championship, as well as future directions, was deliberated. We also examined the performance variables that influence the performance of goalkeepers in the leagues. It is demonstrated that a set of certain performance parameters coupled with goalkeepers' physical characteristics play a key role in determining their performances during matches. In addition, it was demonstrated that the application of data mining and machine learning techniques will continue to be relevant in evaluating goalkeepers' performance.

Keywords Data mining in sports · Goalkeepers performance · European football leagues · Machine learning in sports

8.1 Summary

It has been shown from the findings of the current investigation that several indicators influence the performance of goalkeepers (GKs) across different leagues in the European championship. These indicators consist of anthropometric, technical as well as tactical indicators. It has been demonstrated from findings of the current brief that high penalty savers are essentially taller, older, and heavier in comparison with the low penalty savers. In addition, it was demonstrated that the use of feet could play a role in penalty-saving performance. The high penalty savers commonly used their right legs as a dominant foot, while larger proportions of the low penalty-saving GKs are left-footed, and a few used both feet. Moreover, it has been demonstrated from the study findings that GKs with bigger body sizes and older ages are more likely to keep clean sheets compared to those with lower anthropometrics and younger ages. The study also found that the feet used by the GKs have no association with keeping clean sheets.

The study findings further demonstrated that a set of indicators encompassing accurate passes, key passes accurate, fouls, accurate foot passes from open play, hand passes accurate, accurate passes from set pieces, short passes accurate, medium passes accurate, accurate long passes, and goals conceded could be essential in identifying the performance level of the GKs. Moreover, it has been demonstrated from the study findings that a set of actions encompassing stopped shots, saves without jumping, jumping saves, shots saved with unsuccessful bouncing, shots saved with successful bouncing as well as close-range shots on target are highly dominant for GKs performances in the European football tournament. It has been demonstrated from the study findings that a set of shots and saves encompassing mid-range shots saved, close-range saves, mid-range shots saved as well as long-range shots are vital predictors for GK's goals conceptions.

8.2 Conclusion

To ensure the effective performance of GKs, a range of tactical and technical indicators are crucial. These indicators encompass skills such as accurate interception, precise passing both with feet and hands from open play and set pieces, as well as proficient medium and long-range passes. These aspects play a pivotal role in determining the success of GKs in the European football championship. The findings emphasize that certain body characteristics are linked to successful penalty-saving performance. High penalty-saving GKs tend to be taller, older, and heavier when compared to their low penalty-saving counterparts. Furthermore, the role of foot preference in penalty-saving performance has been explored. High penalty-saving GKs often use their right legs as their dominant foot, whereas a greater proportion of low penalty-saving goalkeepers is left-footed, with a minority using both feet. The study also reveals that GK's performance in terms of maintaining clean sheets is influenced by body size and age. GKs with larger body sizes and older ages are more likely to achieve clean sheets compared to those with smaller anthropometric measurements and younger ages. Interestingly, the research indicates that the feet used by goalkeepers do not show a significant association with their ability to maintain clean sheets.

The findings from the current brief highlight a specific set of actions that strongly influenced GK's performance in the European football tournament. These actions include proficiently stopping shots, making saves without the need for jumping, executing jumping saves, successfully saving shots with controlled bouncing, and effectively dealing with close-range shots on target. These actions significantly contribute to GKs' success on the field. Moreover, the study demonstrates the importance of particular shots and saves in predicting GK success. These encompass mid-range shots saved, close-range saves, and successful handling of long-range shots. Recognizing the significance of these indicators and actions can greatly enhance goalkeepers' effectiveness and contribute to their overall performance in the European football championship.

8.2.1 Current Status and Future Direction

Currently, the performance of GKs in the European football leagues stands as a dynamic and evolving landscape. With advancements in sports science, technology, and tactical understanding, GKs have become integral and versatile players within their teams. The emphasis on technical skills, such as precise distribution and effective ball-playing abilities, has reshaped the traditional role of GKs. Additionally, the incorporation of data analytics has enabled a more comprehensive evaluation of their performance, leading to targeted training and strategic improvements. Looking ahead, the future direction of GK's performance in European football leagues is likely to see an even greater fusion of traditional shot-stopping excellence with modern demands for tactical versatility and distribution skills. The integration of AI and advanced training methodologies is anticipated to refine GKs' abilities further, contributing to a higher level of competition and enhancing the overall quality of play across these leagues.

Printed in the United States
by Baker & Taylor Publisher Services